Sanborn Tenney, Abby Amy Gove Tenney

Natural history of animals

Sanborn Tenney, Abby Amy Gove Tenney

Natural history of animals

ISBN/EAN: 9783337815073

Printed in Europe, USA, Canada, Australia, Japan

Cover: Foto ©ninafisch / pixelio.de

More available books at **www.hansebooks.com**

NATURAL HISTORY OF ANIMALS

BY

SANBORN TENNEY AND ABBY A. TENNEY

ILLUSTRATED WITH FIVE HUNDRED WOOD ENGRAVINGS

CHIEFLY OF NORTH AMERICAN ANIMALS

REVISED EDITION

NEW YORK ·:· CINCINNATI ·:· CHICAGO
AMERICAN BOOK COMPANY

THIS BRIEF ACCOUNT

OF

The Animal Kingdom

IS AFFECTIONATELY DEDICATED

TO THE YOUNG

PREFACE.

THIS little volume contains a brief account of the Animal Kingdom, and it is hoped that it may aid parents and teachers in interesting the young in the delightful and important study of Natural History. As indicated on the title page, it serves the purpose of a key to the Natural History Tablets, but is also complete in itself without the Tablets.

It is proper to add that the engravings are the same, with few exceptions, as those in Tenney's "Manual of Zoölogy," and that those of the Mammals are mainly from Schinz, Audubon and Bachman, and Richardson; of the Birds, mainly from Audubon and Wilson; of the Reptiles and Batrachians, mainly from Holbrook; of the Fishes, from Storer, Holbrook, DeKay, and from nature; of the Insects, from Harris, Emmons, Say, Sanborn, and from nature; of the Crustaceans, mainly from nature and Reports; of the Mollusks, from Binney, Woodward, Gould, Lea, Conrad, and from nature; of the Echinoderms, from nature, Agassiz, and Muller; of the Acalephs, from Agassiz; of the Polyps, from Dana, Milne-Edwards, Verrill, and from nature; and of the Protozoans, mainly from Ehrenberg and Huxley.

Both this volume and the "Elements of Zoology" already announced by Messrs. Scribner and Co., and which will combine the study of the Anatomy and Physiology of Animals with that of Descriptive Zoology, are intended to precede the Manual mentioned above.

VASSAR COLLEGE, POUGHKEEPSIE, N. Y.,
    August, 1866.

In the present edition a few changes have been made, as the volume is no longer to be used in connection with the Natural History Tablets.

    May, 1895.

CONTENTS.

	PAGE.
GENERAL DESCRIPTION OF ANIMALS	11
VERTEBRATES, OR BACKBONED ANIMALS	19
MAMMALS	22
BIRDS	72
REPTILES	114
BATRACHIANS	120
FISHES	124
TUNICATES	139
ARTHROPODS, OR JOINTED ANIMALS	139
INSECTS	139
ARACHNIDS	193
MYRIAPODS	197
CRUSTACEANS	197
MOLLUSKS, OR SOFT-BODIED ANIMALS	203
CEPHALOPODS	205
GASTROPODS	210
HETEROPODS AND PTEROPODS	222
LAMELLIBRANCHIA	223

	PAGE
VERMES, OR WORMS	229
BRACHIOPODS	231
POLYZOA	232
PARASITIC WORMS	233
ECHINODERMS, OR STARFISHES	233
HOLOTHURIANS	233
ECHINOIDEA	234
SEA STARS	238
OPHIURANS	240
CRINOIDS	241
CŒLENTERATES, OR LASSO-THROWERS	242
ACALEPHS	242
POLYPS	252
SPONGES	265
PROTOZOANS	266
CONCLUSION	269
INDEX	273

NATURAL HISTORY OF ANIMALS.

A GENERAL DESCRIPTION OF ANIMALS.

ANIMALS are living beings which feed upon plants, — or, in many cases, upon animals whose food is plants, — and in which the sense of feeling and the power of motion are well developed. The kinds of animals are very numerous, — more numerous than the kinds of trees in the forest and the flowers of the meadows and fields; and they are of all sizes, from those so minute that thousands can sport in a drop of water, to those of large dimensions, like the Horse and the Ox, the Elephant and the Whale; and their forms are as various as their sizes and kinds. But the name *Animal* is given to them all, whatever their size or form, and whether they swim, creep, fly, walk, or run.

Animals are most interesting objects for study, and the child as well as the man is delighted with learning their forms, structure, color, habits, and names, and soon becomes as eager as a naturalist to find a new Bird or a new Butterfly.

Some kinds of animals, as Man, Cattle, Deer, Sheep, Beasts of Prey, Birds, Turtles, Lizards, Snakes, Frogs, and Fishes, have a backbone, and a spinal cord which is enlarged at the forward end into an organ

Fig. 1. — Deer — American Elk.

Fig. 2. — Bird — Duck.

called the brain; and as the backbone is made up of parts called vertebræ, these animals have been named VERTEBRATES. See Figures 1-6.

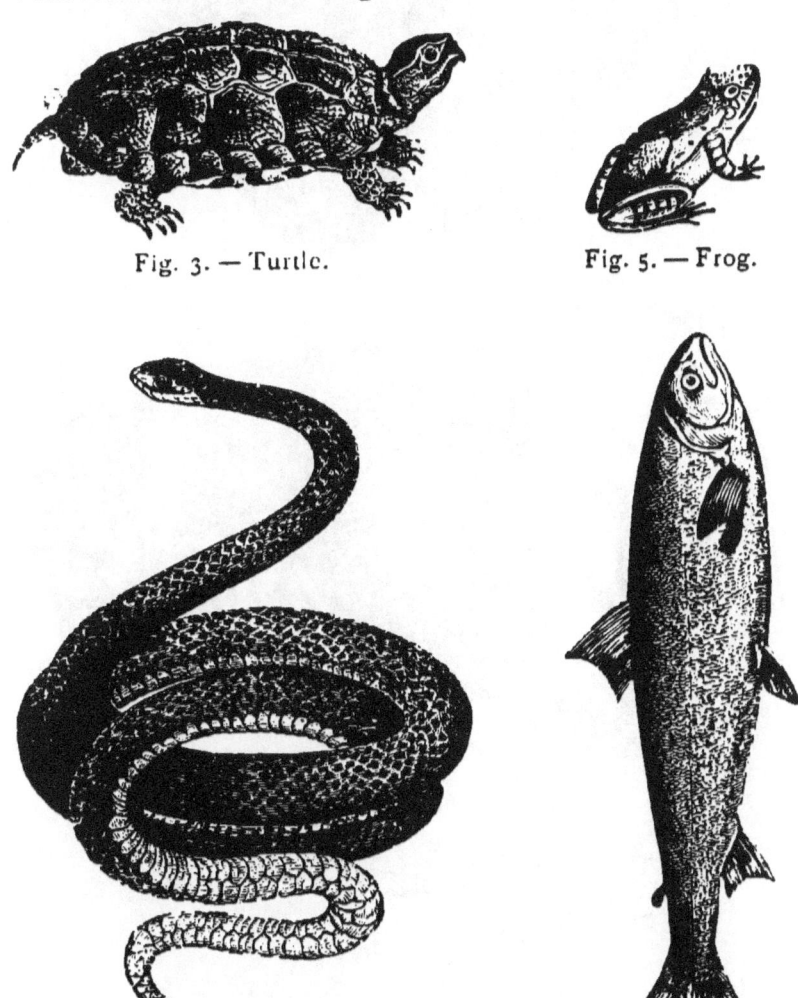

Fig. 3. — Turtle.

Fig. 5. — Frog.

Fig. 4. — Snake.

Fig. 6. — Fish.

Other animals, as Bees, Butterflies, Flies and all other Insects, Spiders, Mites, Crabs, Lobsters, and

14 GENERAL DESCRIPTION OF ANIMALS.

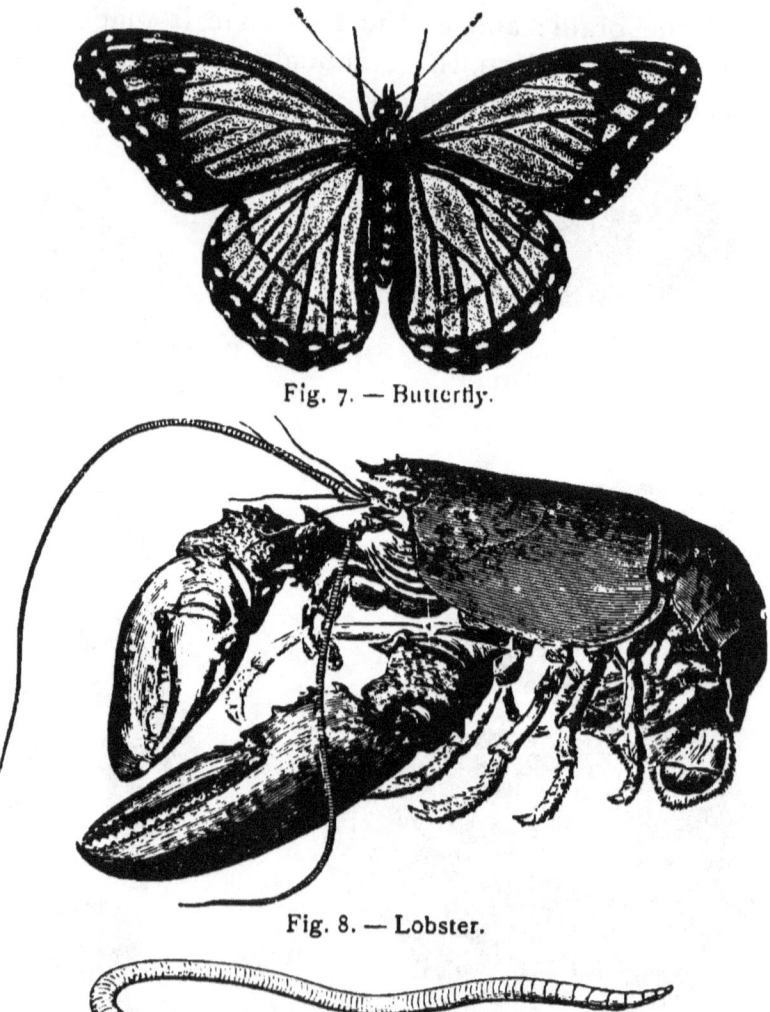

Fig. 7. — Butterfly.

Fig. 8. — Lobster.

Fig. 9. — Earthworm.

Shrimps, are made up of a series of rings, or joints, each bearing a pair of jointed appendages, and hence are called ARTHROPODS from a word which means jointed legs.   See Figures 7 and 8.

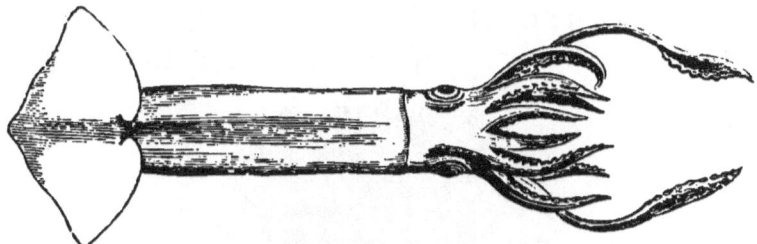

Fig. 10. — Squid.

Fig. 11. — Land Snail.

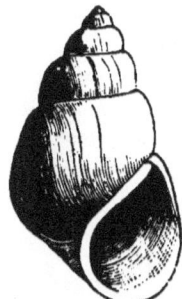

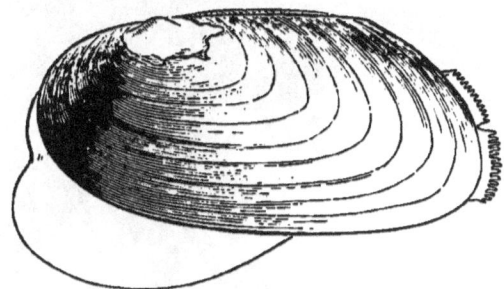

Fig. 12. — Snail Shell.   Fig. 13. — Fresh-water Mussel.

Other kinds of animals, as Squids, Snails, Mussels, Clams, and Oysters, have neither a backbone nor a jointed body; but the whole body is soft, usually with a shell outside, but sometimes without a shell, and they are called MOLLUSKS, from a word which means soft. See Figures 10–13.

16 GENERAL DESCRIPTION OF ANIMALS.

Still other kinds of animals, as Sea Cucumbers, Sea Urchins, Sea Stars, Serpent Stars, and Crinoids are

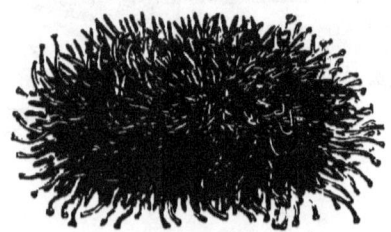

Fig. 14. — Sea Urchin.

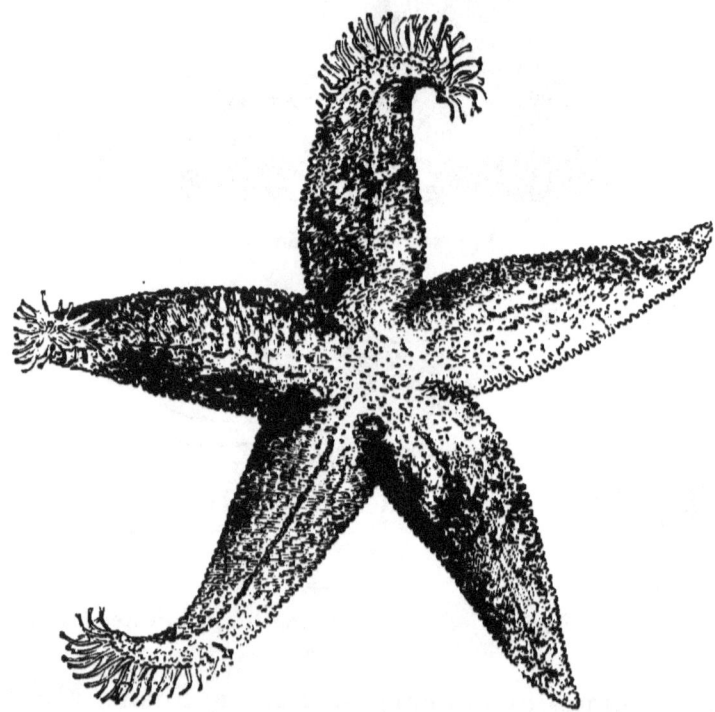

Fig. 15. — Sea Star or Starfish.

star-shaped, or flower-shaped, their parts radiating from a common center or axis. They have a distinct stomach and their skin is usually hardened and covered with

GENERAL DESCRIPTION OF ANIMALS. 17

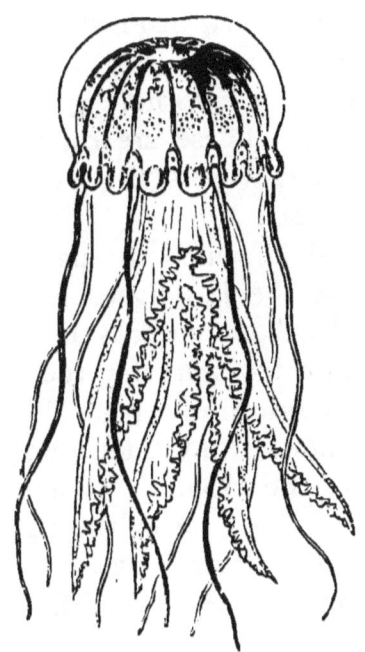

Fig. 17. — Sea Anemone.

Fig. 16. — Jellyfish.

Fig. 18. — Coral Polyps.

Fig. 19. — Coral Polyps.

NAT. HIST. AN. — 2

spines. Hence they are called ECHINODERMS or Hedgehog-skinned. See Figures 14, 15.

A third kind of animals, like the Earthworm (Fig. 9), is jointed, but has no jointed legs. These are called VERMES or Worms. Such are the Sea-worms (Fig. 456), the Hair Worm, and the Vinegar Eel.

There is another sort of animal in which the parts radiate from a center. These are the Jellyfishes, Sea Anemones, and Coral Polyps. In these animals there are always found microscopic lassos for capturing their food. So we may call them the Lasso-throwers. The zoölogists give them the long name CŒLENTERATES, from the fact that the wall of the stomach is not separate from that of the body. See Figures 16-19.

There is a still lower group of animals, related to the Cœlenterates. This is the type of the SPONGES. See Figures 511, 512.

There is a group of still simpler animals, so small that they can be seen only with the aid of the microscope. These are called PROTOZOANS, which word means *first* or *simplest animals*. See Figures 513-520.

VERTEBRATES, OR BACKBONED ANIMALS.

The Vertebrates, as stated on page 11, have a backbone made of parts, each one of which is called a vertebra. This backbone is the most important portion of a bony framework called the skeleton. Upon this skeleton is placed the flesh, and outside of the whole is the skin, which is naked, or covered with hair, fur, feathers, or scales, according to the kind of animal. Within the head is a wonderful organ called the brain, which has a branch called the spinal cord, extending through the body, and contained in a tube formed above the main part of the backbone. From the spinal cord and brain there are little branches, called nerves, which reach to all parts of the head and body. The brain, spinal cord, and nerves are called the nervous system, which

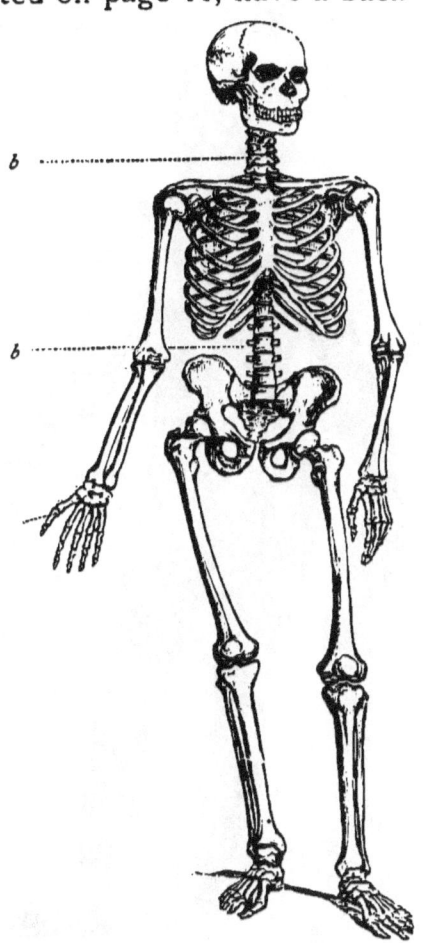

Fig. 20. — Skeleton of the highest Vertebrate — Man.
b, b, backbone.

Fig. 21.

A single vertebra, the round white space showing the place of the spinal cord.

is much the same in its general character in all vertebrates. This system as it appears in Man, the highest vertebrate, is shown in Fig. 22. Besides inclosing the brain and spinal cord, the skeleton protects the organs for breathing and digestion, and supplies the levers by which the muscles move the animal.

As the brain and spinal cord are alike in their position and general outlines in all vertebrates, only differing in size and in degree of perfection, so also are the skeletons of all vertebrates alike in their principal features. The backbone of one, in its position and general outlines, corresponds to that of all the others; so with the head and limbs. The arm of Man, the arm of a Monkey, the wing of a Bat, the leg of a Mole, the leg of a Dog, the paddle of a Seal, the leg of a Sheep, the paddle of a Whale, the wing of a

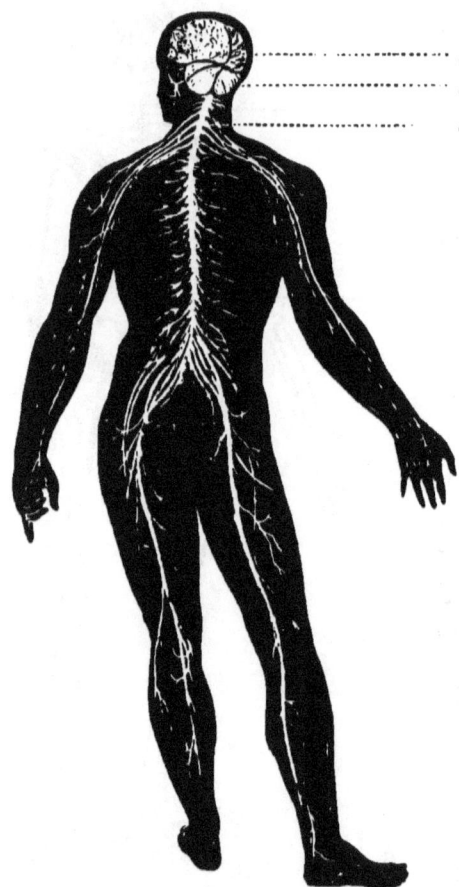

Fig. 22. — Nervous system of the highest Vertebrate — Man.

a, principal brain, called the hemispheres; *b*, smaller brain; *c*, spinal cord giving off its branches of nerves.

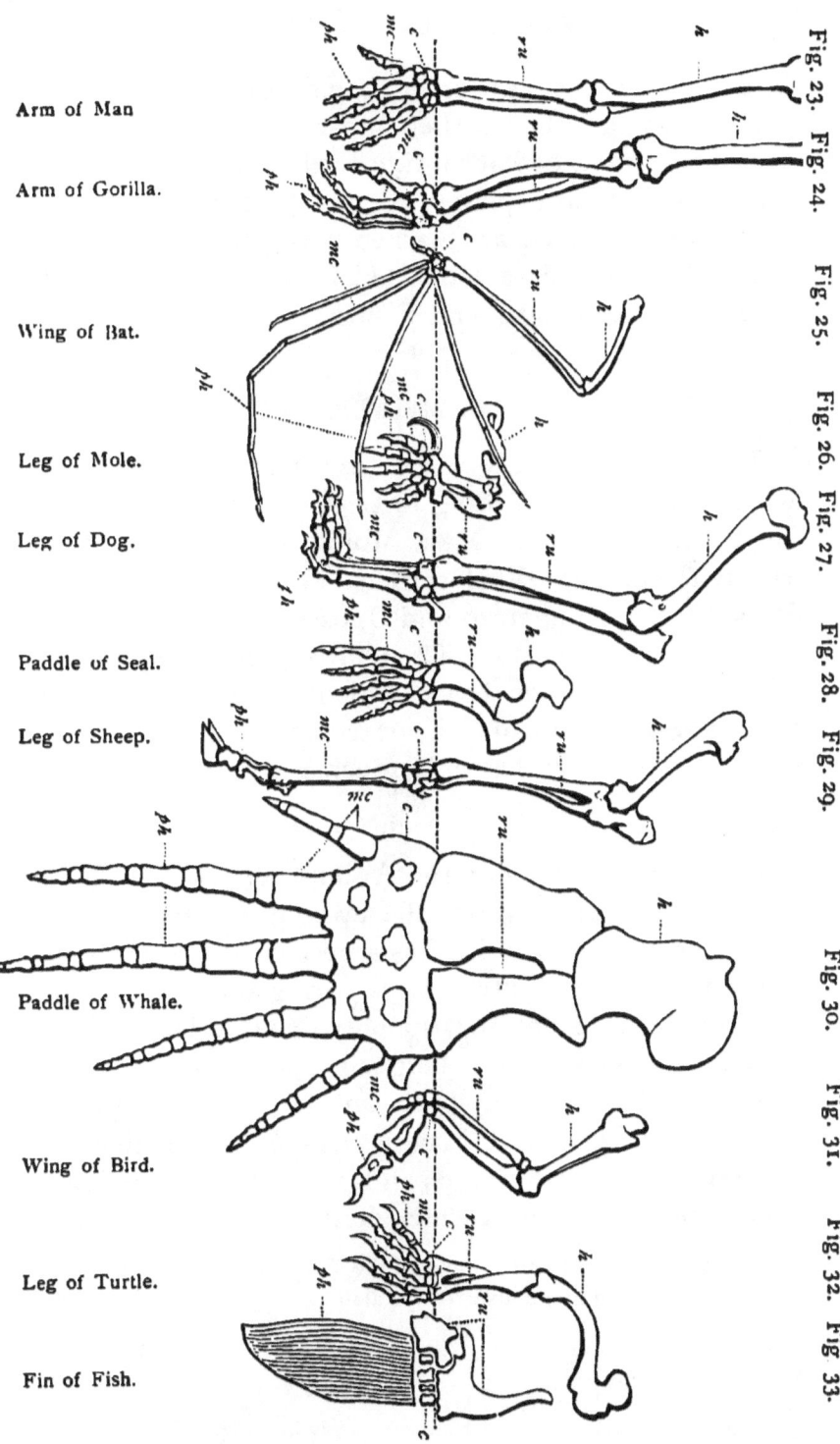

Arm of Man — Fig. 23.
Arm of Gorilla. — Fig. 24.
Wing of Bat. — Fig. 25.
Leg of Mole. — Fig. 26.
Leg of Dog. — Fig. 27.
Paddle of Seal. — Fig. 28.
Leg of Sheep. — Fig. 29.
Paddle of Whale. — Fig. 30.
Wing of Bird. — Fig. 31.
Leg of Turtle. — Fig. 32.
Fin of Fish. — Fig. 33.

Bird, the leg of a Turtle, and the fin of a Fish, correspond to one another in their most important features, each being modified according to the use to which it is put. This is quite plainly seen in Figs. 23-33, where corresponding parts are marked with the same letter.

The Vertebrates are divided into Mammals, Birds, Reptiles, Batrachians, and Fishes.

MAMMALS.

The Mammals are Vertebrates whose skin is covered with hair and which bring forth living young and nourish them with milk. Man, Monkeys, Beasts of Prey, Hoofed Animals, Whales, Bats, Moles, Squirrels and Rats, Sloths, Kangaroos and Opossums, and Duckbills, come under this head. They all breathe air by means of lungs, have warm blood which is sent throughout the body by means of a heart constructed like that of Man, and the neck has only seven vertebræ.

MAN.

Man is at the head of the Animal Kingdom. He is the only animal to whom the upright position is natural; the only one which has a perfect hand; the only one whose forward extremities — arms and hands — are not used for locomotion; the only one that laughs; the only one that speaks a language; his brain is larger than that of almost any other animal,* and he can live in all countries. But Man is also far more than an animal. He has a mind and a soul and can learn much about the things which God has made.

* The brain of the Elephant and of the Whale is larger than that of Man, but the animals themselves are also far larger.

Monkeys, or Quadrumana.

Apes and Monkeys are animals all of whose four feet are hand-like, as the great toe can be shut against the other toes, like a thumb. Hence comes their

Fig. 34.—Chimpanzee.

scientific name, Quadrumana, which means *four-handed*. But though these hands are well adapted for grasping and climbing, they are much inferior to the perfect hand of Man. Some kinds of Ape can stand upright, but not firmly, for the soles of their feet nearly face each other, and cannot be brought flat to the ground like the foot of Man. About eighty kinds of Monkey live in the forests of the warm parts of Asia and Africa, and even more kinds in South

Fig. 35. — Orang-outang.

America. Those of Africa and Asia have thirty-two teeth, their nostrils near together, and their tail, even when present, is not capable of grasping objects. Most of the Monkeys of America have thirty-six teeth, the

MONKEYS.

nostrils far apart, and many of them have the tail capable of grasping objects, and thus of being used in climbing and in picking up objects which cannot be reached by the hand. Monkeys live mainly on the trees, and feed upon fruits, nuts, eggs, and insects. They are mischievous and thievish.

The Chimpanzee of Western Africa is one of the Monkeys having no tail, which are called Apes. Of all its tribe, it is thought to be the most like Man; but the great African Ape, called the Gorilla, is a larger species. Although when in an upright position the Chimpanzee somewhat resembles a human being, its long muzzle and other characters separate it widely even from the lowest tribes of the human family. The Orang-outang is an Ape which inhabits Borneo, and is smaller than the Chimpanzee. The latter may be nearly five feet high. The

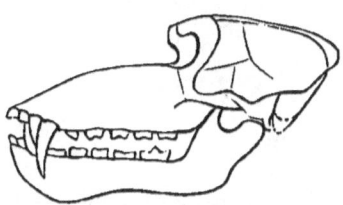

Fig. 36.—Skull of Baboon.

Fig. 37.—Kahau.

Fig. 38.—Spider Monkey.

Kahau of India is about the size of a large dog, and is named from its peculiar cry. The Baboons, often called Dog-headed Monkeys and Mandrills, have a very long muzzle, like that of a Dog, as shown by Fig. 36. They are common in Africa; some of them are very large and ferocious; in appearance they are the ugliest of all the Apes. The Spider Monkey of

Fig. 39. — Marmoset.

Fig. 40. — Lemur.

South America is so called from its long, slender legs. Its long tail is of great aid in climbing. The Marmosets of Brazil are very small and curious Monkeys, with long, soft, and beautifully colored fur.

Fig. 41. — Aye-aye.

The Lemurs, or Makis, are pretty monkey-like animals, most of which live in Madagascar. The tail is quite bushy, and in many respects they much resemble common four-footed animals. The Aye-aye is a curious monkey-like animal, about as large as a Cat, which lives in Madagascar. Its incisor teeth are like those of the Rodents, its middle finger is exceedingly elongated and slender, and its tail is bushy.

Some kinds of Monkeys imitate the actions of men, and their efforts of this sort are often ludicrous.

Flesh-eaters, or Carnivora.

These animals have their teeth and claws very sharp, and they capture and devour other animals for food. In the Cats, the back teeth, or molars, have sharp edges, and those in the two jaws shut by each other like the blades of scissors, and thus cut the flesh into pieces fit for swallowing. In others, like the Bears, the back teeth are fitted for grinding, and such do not live exclusively on flesh.

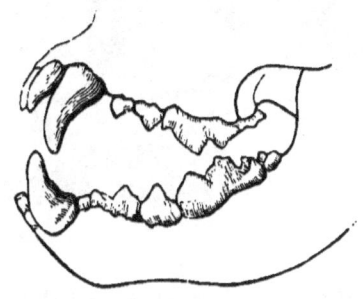

Fig. 42.— Teeth of a Flesh-eater.

Cats, Hyenas, Dogs, Civets, Weasels, Bears, and Seals are the chief Carnivora.

Cats.

Of all the Carnivora the Cats have the keenest senses and the quickest movements, and they are the most rapacious. Their tread is noiseless, the bottoms of their feet being like a cushion; they stealthily approach their prey, and when near enough, seize upon it with a sudden spring. The name *Cat* is given not only to the domestic varieties of this sort, but also to the Lion, Tiger, Panther, Leopard, Puma, Lynx, Jaguar, and Wild Cat. The Lion, Panther, and Leopard inhabit

Africa and Southern Asia, and the Tiger is found in Asia, the first and last being the largest of all the Cat

Fig. 43 — Puma.

tribe. The Puma is found from Canada to Patagonia; it is larger than the largest Dog, and preys upon

Fig. 44.— Canada Lynx.

deer, sheep, hares, and sometimes attacks human beings. It climbs trees, and often lies upon a limb in wait for prey. The Jaguar inhabits South America, and is found as far north as Texas and as far south as Patagonia. The American Wild Cat and Canada Lynx much resemble each other, but the Lynx is the larger, being about three feet long, and has ears tipped with long black hairs. They feed upon small quadrupeds and birds, often pursuing the latter into tree-tops.

Hyenas.

Hyenas live in Africa and Asia, and are about the size of a very large Dog. They live in dens and caves, come forth at night in search of food, feeding mainly on animals which they find dead. They are ferocious and greedy, and have such stout teeth and powerful jaws that they are able to crush the bones of the largest prey, the fragments of which they swallow without masticating them.

Dogs, Wolves, and Foxes.

The Dog is the only animal that has followed man to all parts of the world. The varieties are numerous, and differ from one another greatly in their appearance and habits. Some of the most distinct varieties are the Greyhound, St. Bernard, Newfoundland, Eskimo, Shepherd Dog, Fox Terrier, Deerhound, Bloodhound, Spaniel, Setter, Pointer, Poodle, Terrier, Mastiff, etc. The Dog is noted for its sagacity, courage, and faithfulness.

Wolves are ferocious and greedy animals, about the size of a large Dog. They often hunt in companies or packs, and thus are able to kill animals which singly they could not master. In newly settled parts of the country, they destroy sheep, calves and other animals of the farm. The White and Gray Wolf is found in nearly all the thinly settled regions of North America.

Fig. 45. — American White and Gray Wolf.

The Prairie Wolf is common in the regions west of the Mississippi River.

Foxes are distinguished from all the rest of the Dog family by their pointed muzzle and large bushy tail. They are the most sly and crafty of all animals, contriving to steal turkeys, geese, chickens and whatever they want to eat, and carry them away to their lurking-places in the woods and thickets. They are hunted with hounds, which go in swift pursuit, while the hunter, knowing the habits of the animal, conceals himself in some valley or other locality where the fox will be almost sure to pass, and when it comes near enough shoots him down. But it must be stated that, in many cases, the shrewd movements of the fox deceive both the hunter and the dogs. If captured alive, which rarely happens, and struck while it is in a situation from which it cannot escape, the fox feigns itself dead, though unhurt, and when its captor is off his guard, will jump up and run away.

Civets.

Civets are about the size of the house Cat, and with one exception belong to the Old World. The Civet of Texas and California is of a grayish color, its tail being white with black rings. It lives upon the trees, is lively and playful, and, though shy, is easily tamed, and sometimes kept as a pet.

Fig. 46. — Civet.

Fishers, Martens or Sables, Weasels, Otters, etc.

These animals have, in most cases, a slender body, and long soft fur, especially in winter. They are quick in movement, and destructive to other small animals.

The American Fisher is about the size of a Cat, but with a much more slender body, and is nearly black. The American Sable, or Pine Marten, of the Northern States and Canada, is much smaller than the Fisher, of

Fig. 47. — Weasel. Fig. 48. — American Sable.

a brownish-yellow color, and is celebrated for its beautiful and valuable fur, which is generally called the Hudson Bay sable. The fur known as the Russian sable comes from a very similar animal which lives in Siberia. The Pine Marten delights in dense woods, where it pursues and captures hares, birds, and squirrels, swiftly following the latter even among the tree-tops. Its retreats, especially in winter, are hollow trees, and it is often seen by the hunter sitting with the head just out of its hole. If shot while in this position, it falls back into the hole and is lost; so the hunter, knowing its habits, walks slowly around the tree; the sable comes out to gratify its curiosity by a look at the hunter, and is then shot and falls to the ground. More than a hundred thousand skins of this animal have been collected in northern North America in a single year.

True Weasels vary from five inches to a foot in length, and are generally brown in summer and white in winter, the tail tipped with black. There are half a dozen kinds in North America. The fur known as ermine is furnished by the Weasels, the most valuable coming from Siberia. Weasels are generally bold, courageous, and extremely bloodthirsty, eagerly attacking animals much larger than themselves. They destroy rats and birds, and commit great havoc among poultry, a single individual having been known to kill fifty chickens in one night and the evening of the following day, and to kill several chickens in a coop near which a man was standing.

Fig. 49 — Mink.

Minks are about a foot and a half long to the tail, and are dark brown or black. They are found about ponds and streams, and their fur is very beautiful, and is often sold under the name of American sable.

FLESH-EATERS.   33

The Wolverine, found in the Northern States and Canada, and in the northern part of Europe and Asia,

Fig. 50. — Wolverine.

is about three feet long, of a dark color, and is very powerful and ferocious when attacked. It is very troublesome to sable hunters, breaking down their wooden traps, and eating the bait and game. It is so shrewd that it scarcely ever enters the trap, and hence is not often caught.

Fig. 51. — American Otter.

34 VERTEBRATES: MAMMALS.

Otters live in and about the water, and feed upon fish. They are sportive in their disposition, and amuse themselves by "sliding down hill." Selecting a steep bank of a river, they slide head foremost into the water, and repeat the operation many times, apparently with delight. Otters are three or four feet long from the nose to the tip of the tail, the color dark brown, and the fur is of two kinds, one short, fine, and thick, the other long, coarse, and scattered. When taken young, Otters are easily tamed, and become so familiar that they will lie in the lap like a cat.

Fig. 52. — Skunk.

Skunks are found only in America, and are notorious on account of their disagreeable odor. They are a foot and a half long to the tail, and the color is black and white. They live in burrows, and seek their food at night, eating beetles and other small insects, and eggs. Since their food consists so largely of insects, they are useful to the farmer. Their fur is also of commercial value.

Fig. 53. — American Badger.

Fig. 54. — Grizzly Bear.

The Badger of western North America is about two feet long, with a stout body and short tail, and its color is gray. The hair is long, extending on the hind part of the body so as nearly to conceal the tail. Badgers live in burrows, and dig with astonishing rapidity.

Bears and Raccoons.

Bears and others of this family walk on the sole of the foot. They feed upon flesh, berries, and roots.

The Raccoon of the United States is about as large as a middle-sized Dog, with a thick body, looking somewhat like a small Bear with a long tail; the color is grayish, and the tail is ringed with black and dingy white.

Fig. 55. — Raccoon.

Bears are very large. The Grizzly, of the Rocky Mountains, is six or eight feet in length, and weighs in some cases eight hundred pounds, and the nails or claws are six inches long. It is the most powerful animal in America, and when wounded is very dangerous to the hunter. It has been seen to drag away a large bison, after killing it. The Black Bear of the Northern States is much smaller than the Grizzly and less ferocious, seldom attacking men when not molested; but if disturbed when accompanied by its cubs, it fights very savagely.

Seals and the Walrus.

The Seals and the Walrus live in the sea, but often come upon the rocks and ice-banks to lie in the sunshine. The head of the Seal much resembles that of a Dog, and its eyes are beautiful and intelligent in appearance. When taken young, Seals are easily tamed, and become attached and obedient to those who feed

Fig. 56.— Seal.

them, coming at call and performing curious feats according to their master's directions. Some years ago, in a large tank of sea water in the Aquarial Gardens at Boston were two seals called "Ned" and "Fanny," which were so tame that they would come to the keeper at call and allow him to handle them, would shoulder a miniature musket, turn the crank of a hand organ, shake hands with the bystanders, and "Ned," especially, would even "throw a kiss" to the ladies. Seals feed upon fish, and always eat in the water. They are from three to twenty feet long.

The Walrus has a body as large as the largest Ox, and is covered with short brown hair. Two of its upper teeth, the canines, or eyeteeth, in the male grow to be tusks two feet long. These tusks assist in climbing upon the ice-banks, serve as a means of defense, and aid in securing food. The Walrus is found in the Arctic Ocean. Their skins, oil, and ivory are valuable.

Hoofed Animals, or Ungulates.

These are Mammals which feed wholly upon vegetation, and which have hoofed feet, and use their limbs only for standing, walking, and running. Some of them, as the Hog, Deer, Antelopes, Sheep, Goats, and Oxen, have the foot divided or cleft, forming an even number of toes. Most Ungulates of this sort chew the cud, and from the latter fact are known as Ruminants, a name which means *cud-chewers*. Others, as the Horse, Ass, and Rhinoceros, have only one toe or an odd number of toes. There are thus two groups of Hoofed Mammals, the odd-toed and even-toed Ungulates. Most of the domestic animals belong to the Ungulates.

Hogs.

The Hog has four toes,—although only two are used in walking,—a long snout, coarse bristles, a simple stomach, and teeth fitted for a mixed diet. There are incisor teeth in both jaws; the grinders are capped with rounded elevations. The purely herbivorous Cattle and Horses have ridges of enamel on the grinders.

There are numerous kinds of Hogs, more than fifteen having been described, mostly from the old world. The Wild Boar of Europe is the race from which our Domestic Hog has sprung. Perhaps other species have been tamed in other parts of the world. Many naturalists think that the Hog of China and Eastern Asia came from another species. The crossing of this form with our native Hogs has given rise to many of the best breeds.

The wild Hogs of America are quite different from the Wild Boar, and are small animals, called Peccaries. They are chiefly found in South America, but one kind is found as far north as northern Texas.

Hippopotami.

The Hippopotamus is a huge hog-like animal, living in the rivers of Africa. It measures as much as twelve feet in length. It lives in herds of twenty to forty individuals in the beds and near the banks of rivers, where it finds its food. This is chiefly grass and water plants, of which it consumes an immense quantity, as its stomach can hold five or six bushels. In places inhabited by man it often does great harm to the fields and gardens, whose products it prefers to the wild vegetation.

Cud-chewers, or Ruminants.

The remainder of the even-toed Ungulates are called Ruminants, from the fact that they chew the cud. The stomach has four divisions. The food goes into the first of these, the paunch, when first swallowed, and is afterwards brought up and chewed again. When swallowed the second time it goes into the true stomach, where it is digested. To this group belong Camels, Deer, Oxen, Sheep, Goats, and Antelopes.

Camels and Llamas.

The Camel is a native of Central and Southern Asia, and, from the earliest times, has rendered such important services to the inhabitants of the East in carrying merchandise across the deserts, that it has been called the "ship of the desert." Its feet are fitted for traveling in the sand, being covered with horny pads rather than with hoofs. Its strength and power of endurance are very great, it can live on the coarsest and most scanty vegetation, and travel for days without drinking. It can carry from five hundred to one thousand pounds.

Fig. 57. — Llama.

The Camel is larger than the Horse, and stands very high. There are two kinds, — one with two large humps upon the back, and the other with only one hump. Both live in the Old World and are found only as domesticated animals.

The Llamas inhabit the Andes of South America, are much smaller than the Camel, being only four or five feet high, and have no hump. They live in herds, and are tamed and used as beasts of burden. The Alpaca is a variety of Llama with long woolly hair, which furnishes material for valuable fabrics. This is also a domestic animal; there are also wild forms known as the Guanaco and the Vicuña.

Deer.

The Moose, Reindeer, Deer, and Elk all belong to the Deer family. The males have solid, bony horns called antlers, which they shed once a year, new and larger ones growing to take the places of those which have been shed.

Fig. 58. — Moose.

The Moose is the largest of all the Deer kind. It is as large as a Horse, and has an exceedingly long head, large flattened horns, and very long legs. It travels with an awkward gait, but with great speed, easily making its way through deep snows, bushes, over brush-heaps, fallen trees, fences, and whatever obstructions lie in its path. It was quite common in some parts of

Maine, northern New York, and Canada, but is rapidly disappearing. The color is grayish brown.

The Reindeer is a much smaller animal than the Moose, being about five feet long and three feet high. It has become celebrated for the services it renders the Laplanders, who keep large herds of Reindeer, and use them for beasts of burden and for drawing their sledges. Their milk and flesh are good for food, and their skins are used for clothing. They are very hardy animals,

Fig. 59. — American Reindeer, or Caribou.

and subsist on the coarsest fare, eating the tender portions of shrubs in summer, and in winter scraping the snow from the ground and feeding upon the "reindeer moss." The American Reindeer, or Caribou, of Maine and Canada, and other northern parts of North America, is by some thought to be of the same kind as the one found in Lapland. Unlike their relatives, both the male and the female Reindeer have horns.

The American Elk, or Wapiti, is another kind of Deer which lives in the wooded regions of the northern parts of North America. It is nearly as large as the Moose, and has horns five or six feet long, and very much branched.

Fig. 60. — American Elk, or Wapiti.

The Common Deer, of the wild regions of the United States, is one of the most beautiful and graceful of all its family. It is very timid, and, when alarmed, bounds swiftly away. It is about the size of a Sheep, but with a much more slender body and much longer legs. It is hunted in the autumn and winter, and great numbers are sent to the markets. Its flesh is called venison, and is highly prized for food.

Fig. 61. — Common, or Virginia, Deer.

Fig. 62. — Musk Deer.

The Musk Deer inhabits Thibet, and is much smaller than the Common Deer, and has no horns. In each side of the upper jaw are long canine or eyeteeth, like tusks. The musk used in making perfumery is furnished by this animal. It is contained in a pouch on the under side of the body.

Antelopes.

Antelopes are found in Europe, Asia, Africa, and North America, but are most numerous in southern

Africa, where there are many kinds, and where herds of thousands are sometimes seen together. Their horns are hollow, composed of horn, and are variously wrinkled and curved. Antelopes vary in size from those as small as a Deer to those as large as a Horse. Most of the Antelopes belong to Africa.

The Pronghorn Antelope, of the Rocky Mountains, is larger than a Sheep, with much longer neck and legs. Its hair is coarse and thick. It gets its name from the

Fig. 63. — Pronghorn Antelope.

prong, or branch, on each horn. This animal was found at times in large numbers, herds of a thousand and more having often been seen. The progress of settlement has exterminated the antelope over much of the region which it once inhabited and in which it was so abundant. It sheds its horns annually, and is the only Antelope which does so.

The Mountain Goat of the Rocky Mountains, is an Antelope, and not a true Goat, as one would suppose from its name. It is, however, nearly allied to the Goats. It is entirely white, except its horns and hoofs, which are black. Its fleece is long and very fine, being equal in quality to that of the celebrated Cashmere Goat. It inhabits the lofty peaks of the mountains, frequenting the steepest places.

Fig. 64. — Rocky Mountain Goat.

The Gazelle, of Africa and Asia, is about the size of a small Deer, and is celebrated for its beautiful and graceful form, and for its large, dark, and lustrous

Fig. 65. — Gazelle.

Fig. 66. — Chamois.

eyes. The Orientals, or inhabitants of the East, compliment a lady by comparing her eyes to those of the Gazelle. When taken young, though wild and timid, it is easily tamed, and becomes a great favorite.

The Chamois, of the high mountains of western Europe, is about the size of a goat, of a dark brown color, and its horns, towards the summit, are bent backwards like a hook. It is very shy, and on the slightest alarm bounds swiftly away over rocks and glaciers, along dizzy heights, where it would seem no animal could get a foothold, often leaping upon a rock just large enough to receive its four feet placed together.

Sheep and Goats.

Sheep have the horns angular and directed back-

Fig. 67. — Mountain Sheep, or Big-horn.

ward, then spirally curved forward, and yellowish-brown in color. The Mountain Sheep, or Big-horn, of the Rocky Mountains, is much larger than the Domestic Sheep, and has very large horns. The hair is of a gray color and very coarse.

Goats have the horns directed upward and backward, and the chin usually has a long beard. The wild kinds live upon high and rugged mountains, most of them in Asia. The Wild Goat of Persia is supposed to be the parent of the common Domestic Goat. The Cashmere Goat of Thibet is celebrated for its fine wool. Its hair is long and silky; under it is a delicate gray wool, of which the costly Cashmere shawls are made.

The Musk Ox.

The Musk Ox, of Arctic America, is of the size of a small Cow, with very long, dark-brown, silky hair. It

Fig. 68. — Musk Ox.

lives in herds, sometimes numbering nearly one hundred. It feeds upon grass in the mild season, and in winter upon mosses and lichens, from the steep sides of hills blown bare by the winds, and up which it climbs with agility. In spite of its name, it is more nearly related to the Sheep than to the true Ox.

The Bison, or American Buffalo.

The Bison, or Buffalo, of the western plains, is the largest quadruped of America, being of the size of a large Ox. It is covered with thick dark hair, that about

Fig. 69. — Bison, or American Buffalo.

the head and shoulders being long and shaggy. At the time of the discovery of America, the Buffalo was found even to the shores of the Atlantic. It was not uncommon to see the prairies covered with Buffaloes as far as the eye could reach; travelers have passed through herds of them for days in succession, with scarcely any apparent lessening of their numbers. Their paths resembled traveled roads; and as their routes, in most cases, extended in a straight line from one convenient crossing-place of a river or ravine to another, taking springs or streams in their course, they used to serve as highways of travel to the explorers of the plains.

The Buffalo is now nearly extinct. At present there are only a few hundred alive, most of them in Yellowstone Park. As soon as the railways were built across the continent they were rapidly exterminated, being hunted for their skins.

Oxen.

The true Ox is known only as a domesticated animal, although there are some half-wild herds in certain parks in England which may represent the wild stock from which the Domestic Cattle came. In the time of Cæsar, wild Cattle were abundant in the forests of Europe.

Perhaps the Zebu, or "Sacred Cattle of India," had a different source, as there are several wild forms of Cattle in Asia, nearly related to the Domestic Ox.

Odd-toed Ungulates.

The Horse, Ass, and Zebra.

In the Horse and its allies, the weight of the body is carried by the middle toe, the third in each foot. The second and fourth toes are represented by the splint bones. There are incisor teeth in each jaw and the stomach is simple.

The Horse has been found in a wild state in the high plateaus of Central Asia, and perhaps this form represents the ancestor of the Domestic Horse. The wild Horses of America are the offspring of domesticated animals. A great amount of variation has been produced by man in the size, structure and habits of the

Horse. The size of different breeds ranges from the Shetland Pony, weighing hardly 150 pounds, to the Draught Horse, weighing nearly a ton. Equally great are differences in speed and bodily proportions.

The Ass is a native of the Old World, where it is still found wild in Asia and northern Africa. It was one of the earliest animals to be tamed by man.

The striped Zebras, of which there are three or four kinds, are all inhabitants of Africa. None of them have ever been domesticated.

The Rhinoceros.

The Rhinoceros is readily recognized by the horn placed on the nose. This horn differs in structure from those of either the Deer or the Ox. In the Deer, the horn or antler is made of true bone and is shed and replaced annually. In the Ox, the horn is developed from the skin, is hollow, and is borne on a bony core. It is not shed. The horn of the Rhinoceros is also permanent, but is composed of hair-like fibers fastened together. In some kinds of Rhinoceros there are two horns placed one behind the other. The forward horn may be as much as four feet long.

The animal has three toes on each foot. The skin is very thick and is arranged in shield-like folds in the Indian Rhinoceros. In the African forms this arrangement is not found, though the skin is exceedingly dense and is used by the natives for shields.

The Rhinoceros is the largest animal after the Elephant. Some are over fifteen feet long and six or more feet high. Even these, however, probably weigh but little more than the Hippopotamus.

In former geological times, the Rhinoceros lived in Europe and northern Asia. It was adapted to a cold climate by a thick coat of long hair and wool. Its body has been found preserved in the frozen soil of Siberia.

Tapirs.

The Tapirs are found in Central and South America and in the Malay Islands. They are large, heavy animals living in swamps and wet places, feeding on vegetation. They conceal themselves by day. They are hunted for the sake of their flesh and hides.

Elephants, or Proboscidea.

The Elephant is the largest land animal, being sometimes nearly ten feet in height and weighing over 8,000 pounds. The name of the group comes from the proboscis, or trunk, which is the greatly elongated nose. The head is very large and the neck is short, so that all food and water are obtained by the trunk. There are five toes on each foot, although there are not so many hoofs. The Asiatic Elephant has four hoofs on each foot, while the African form has only three on the hind feet.

The tusks of the Elephant are its incisor teeth. Most of the ivory of commerce comes from Africa, where perhaps one hundred thousand Elephants are annually killed for their ivory. The tusks of the African Elephant are much larger than those of the Asiatic, and are borne by both males and females, while only the male of the Asiatic Elephant has tusks.

The Asiatic Elephant is further distinguished from the African, by the fact that the forehead is concave and the ears far smaller than those of its African relative. The Asiatic form has been tamed from very early times, and many stories are told of its strength and sagacity. The African Elephant was tamed by the ancient Egyptians, but no nation at present employs it in domestic service. The Elephant rarely breeds in captivity, and the supply is kept up by annual hunts for new animals, which are soon tamed and set to work.

The Elephants are now far less widely distributed over the world than in former times. The Mastodon and Elephant were formerly found both in America and Europe, even after the coming of man to those regions. In France a piece of the tusk of the Elephant has been found with a picture scratched upon it of the Elephant, drawn by some prehistoric artist.

Whales, or Cetaceans.

These Mammals live in the water, have their limbs paddle-like and fitted for swimming, and their whole appearance is fishlike; but they are true mammals, nourishing their young with milk, breathing air for which they come often to the surface of the water, and their blood is warm. Most of them are large, some being the largest of living animals. They are covered with a smooth skin. They breathe through a hole, or holes, on the top of the back part of the head, through which some kinds spout spray to great heights.

Right and Sperm Whales.

The Greenland, or Right, Whale attains the length of sixty or seventy feet. It has no real teeth, but in the upper jaw are rows of upright horny plates, called whalebone, which are fringed on their inner edges. Its food is small marine animals. Swimming through

Fig. 70. — Skull of the Right Whale, showing the whalebone.

schools of these, the Whale takes millions into its mouth at once, and strains off the water through the whalebone plates, leaving the food in its mouth. This Whale supplies the world with whalebone, and also furnishes more oil than any other. Its home is in cool and frigid seas.

The great Sperm Whale, of the warm parts of the ocean, is fully equal to the Right Whale in size. The upper jaw has neither teeth nor whalebone, but the

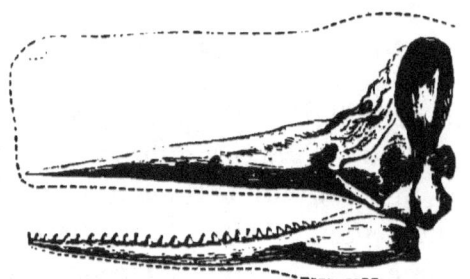

Fig. 71. — Head of the Sperm Whale.

lower has teeth. In the upper portion of the head there are cavities filled with oil, which hardens when cool and is known as spermaceti. The body yields sperm oil. Ambergris, a substance used by chemists in making perfumery, is found in the intestines of this Whale.

The spouting, or blowing, is different in these two Whales; for the Right Whale has two blow-holes on the top of the head, and the spout goes straight up or

WHALES.

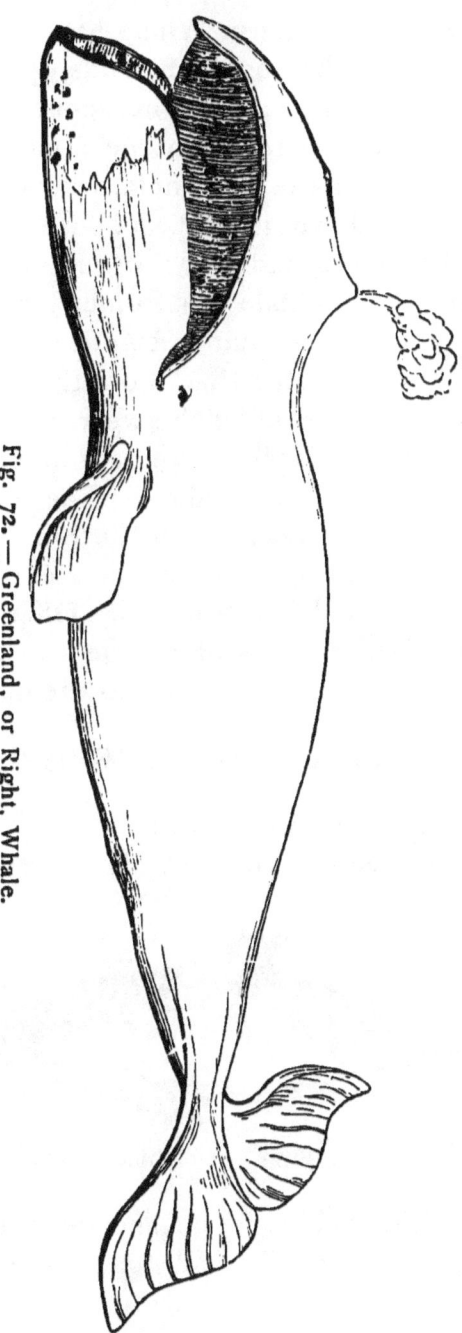

Fig. 72.—Greenland, or Right, Whale.

turns backward; the Sperm Whale has only one blowhole at the end of the nose, and sends up a low bushy spout, which turns forward. The spout is caused by the condensation of water from the lungs and by the water which lies in the blow-hole, which is violently forced out and blown into spray. No Whale takes water into the mouth and blows it out at the nostrils.

The chase of the Whale was formerly a very important industry in which many American vessels were engaged. It is estimated that more than a quarter of a million Whales were killed between 1835 and 1878. The Whales are now far less abundant and the demand for their products is far smaller. The whalebone is probably now more valuable than the oil. Nowadays, the Whale is hunted with the steam whaler and killed by harpoon guns and bomb lances. Off the coast of Norway, the smaller kinds of Whale are hunted with steam tugs, which tow their catch to the harbor.

Dolphins, Porpoises, and the White Whale.

These animals live in herds, and prey upon fishes. The Common Dolphin is about eight feet long, black

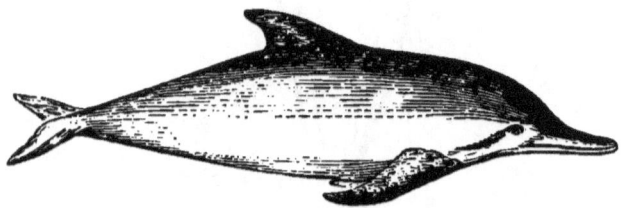

Fig. 73. — Dolphin.

above and white below. The ancients believed this animal to be very docile and fond of music. The

White Whale lives in the northern seas, and is from ten to twenty feet long. It often ascends rivers, and

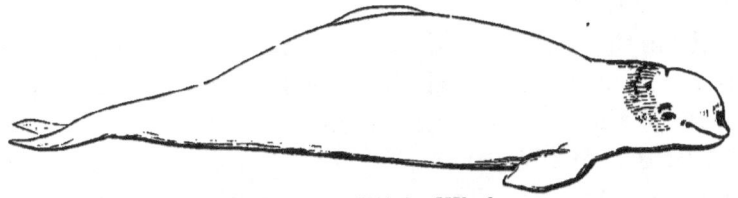

Fig. 74. — White Whale.

is frequently seen in the St. Lawrence. One about ten feet long, was kept for two years in the Aquarial Gardens in Boston. He was quite docile, knew his keeper, would take food from his hand, was trained to a harness, and drew a car prepared for the purpose.

The Mammals already described are mostly of large size; we now come to the smaller ones.

Bats, or Cheiroptera.

Bats are animals which have a thin skin reaching from the arm to the hind legs, and extending to the ends of their long fingers. By means of this skin they can fly as easily as birds, and their flight is noiseless

Fig. 75. — Hoary Bat.

and rapid. The body is covered with soft fur. Their eyes are very small, their ears large, and the thumb has a sharp hook. In the daytime they stay in caves, hollow trees, or other dark places, hanging by their hooks, or by the sharp claws of their hind feet. Bats can fly through the most winding and crooked passages without harm, even after their eyes have been destroyed. Some of the larger Bats of the East Indies eat fruits and birds, but most kinds feed upon insects, which they are catching when we see them flitting in the dusk of evening. The Red, and the Hoary Bat, three or four inches long, are common in North America.

Insect-eaters, or Insectivora.

The Insect-eaters include the Shrews, Moles, and Hedgehogs. Many naturalists also place here the Gale-

Fig. 76. — Galeopithecus.

Fig. 77. — Teeth of an Insect-eater.

opithecus, a curious batlike animal found on trees in the Indian Archipelago. The Insectivora sleep during the day, and go forth at night in search of food. In cool regions, many of them sleep all winter.

Shrews.

Shrews are little mouselike animals, — some of them smaller than the smallest Mice, — with a long and tapering head and soft silky fur. They live under rubbish or dig homes in the ground, are very quarrelsome, and

Fig. 78. — Thompson's Shrew.

Fig. 79. — Water Shrew.

if two are confined together the weaker is soon killed. North America has more than a dozen kinds.

Moles.

Moles have a stout, thick body; short, strong legs; a short tail; and very large fore feet fitted for digging. They feed on earthworms and insects. Their long burrows are their hunting grounds, which they range in search of food. Their eyes are very small, and their fur is soft, thick, and velvet-like. The Shrew Mole of North America is of the size of a very large

Fig. 80. — Shrew Mole.

Mouse, and its eyes are so small that many suppose it to be blind. The hole for the eye is only about the size of a hair, and the eyeballs are smaller than a mustard seed. The Star-nosed Mole is about the size of

Fig. 81. — Nose of Star-nosed Mole.

Fig. 82. — Skull of Star-nosed Mole.

the Shrew Mole, and is so named from the form of the end of the nose, which is star-shaped.

Hedgehogs.

These animals are short and thick, and the back is covered with spines. When alarmed, they take the

Fig. 83. — Madagascar Hedgehog, or Tenrec.

form of a ball, presenting the spines in every direction, to ward off attacks. They sleep during the day in concealed places, and come forth at night to feed upon

Fig. 84. — European Hedgehog.

insects, fruits, and roots. In cold climates they sleep all winter. They live in the Old World and are all small, the European Hedgehog being about ten inches long. The so-called American Hedgehog is a Porcupine.

GNAWERS, OR RODENTS.

The Rodents are readily known by their teeth. In each jaw they have the two front ones chisel-shaped, and between these and the grinders there is a wide space without teeth. The cutting teeth have the hard enamel only on the front of the tooth; thus they wear in such a manner that the more they are used the sharper they become, and they grow at the base as fast as they wear away at the top. Hundreds of Rodents are known, most of which are small.

Fig. 85. — Skull of a Rodent.

The Rodents include the Squirrels, Gophers, Woodchucks, Rats and Mice, Porcupines, Hares, etc.

Squirrels.

Squirrels are small and very pretty animals, with large bright eyes, long ears, divided upper lip, and long bushy tail. They are lightly built, agile, and live in trees, feeding upon fruits and nuts. There are about one hundred kinds in North America, many new kinds having recently been discovered. The most prominent are the large Fox Squirrels of the Middle, Southern, and Western States, and the well-known Gray, Red, and Flying Squirrels found over a large part of the United States. The Gray Squirrels are noted for their occasional extensive migrations. Assembling in large numbers, they cross the country, swimming

Fig. 86. — Gray Squirrel.

rivers, and turning aside for no obstacle. Gray squirrels occur of every shade from gray to jet black.

The Red Squirrel is seen at all seasons and in all weathers. In the northern forests, the deepest snows of winter are soon covered with its tracks, and penetrated by holes bored to find the cones of spruce and pine, and the nuts scattered or hidden beneath. It often sits for hours upon a stump or limb of a tree, and, holding a cone or nut in its fore paws, gnaws it briskly till it gets all the food it contains. If disturbed while upon the ground, this squirrel runs up the nearest tree, leaping from branch to branch, and tree to tree, soon passing out of sight. Sometimes, when startled, it commences chattering with great fury, and leaping about as if in defiance of the intruder.

The Flying Squirrels have a thin skin, or membrane, covered with fur, which extends along the sides of the body between the fore and hind legs, and which, when spread out, serves as a support in leaping from tree to tree, and enables them to perform a sort of flight. They are nocturnal, and therefore not often seen. Their nests are made in the hollows of trees, where large companies often live together. The Common Flying Squirrel of the United States is about five inches long, and the fur is soft, silky, and yellowish brown. It is quite easily tamed, and, being gentle and very beautiful, makes a pleasant pet.

Fig. 87.— Flying Squirrel.

The Striped Squirrels have cheek-pouches, in which they carry grain and nuts to their holes, and they have a shorter and less bushy tail than the others. The

Common Striped Squirrel, or Chipmunk, is about five inches long to the tail, and the color is yellowish gray with five black stripes on the back and sides. In au-

Fig. 88. — Striped Squirrel, or Chipmunk.

tumn the Chipmunks may be seen with their cheek-pouches full of nuts or grain, which they store up for winter, at which time they remain in their holes.

Gophers.

The Striped Gopher, of Michigan and westward, is a very beautiful animal, about the size of the Red Squirrel, of a dark brown color, with light lines and rows of light spots. It lives in burrows, and when alarmed pops into its hole with a chirp. The Prairie Dog is larger than the Striped Gopher, appearing somewhat like a small Woodchuck. It utters a sharp chirp, called bark-

Fig. 89. — Striped Gopher.

Fig. 90. — Prairie Dog.

ing; hence its name. It lives in burrows, and large numbers are found together, forming communities called *dog towns*. Before each hole is a little hill of earth, upon which the Prairie Dog sits on the lookout for intruders. At the slightest alarm it dives into its hole, but soon reappears. Their holes are also the home of the Burrowing Owls and Rattlesnakes.

The Pocket Gopher, Pouched Rat, or Geomys, of the prairies of the Western States, is nine or ten inches long, with large front teeth, strong fore feet, and a short tail. Opening on the outside of the mouth are large

Fig. 91.— Pocket Gopher.

cheek-pouches, which reach back even to the shoulders; and these pouches are lined with fur, and are entirely different from the much smaller cheek-pouches of the Striped Gopher, which open within the mouth. The Pocket Gopher throws up a mound of earth which, in some instances, is ten feet in diameter, and two feet high; and within this mound is its nest, where it rears its young. From the mound it digs numerous galleries in different directions, one or two feet below the surface of the ground. It uses its curious pouches for carrying food, and for carrying away the earth which it removes in digging its galleries. Coming to the surface with its pouches full of earth, it empties them so quickly as to puzzle the looker-on, and instantly retreats into its hole. Pocket Gophers feed mainly upon the roots of plants. They are savage and offer battle to man. If two are placed together, they instantly attack each other, and the stronger eats up the weaker.

Beavers.

Beavers are about three feet long to the tail, and are the largest of the Rodents, excepting an animal called the Capybara which lives about the rivers of South America. Beavers have a flat, scaly tail, and are wholly aquatic in their habits. Their food is chiefly bark and aquatic plants. Their teeth are very sharp and

Fig. 92 — American Beaver.

powerful, enabling them to gnaw down trees of the hardest wood. Beavers prefer running water, that the wood which they cut may be carried to the desired spot. They keep the water at a given height by dams, built of trees and branches mixed with stones and mud; winter houses are built of the same materials. Each house consists of two stories; the upper is above water and dry, and serves as a shelter; the lower is beneath the water, and contains stores of barks and roots; the only opening is beneath the water. The Beaver is reddish-brown; the fur is soft and fine. It lives in unsettled parts of North America, but is nearly extinct.

Rats and Mice.

There are more than three hundred kinds of these animals, all of which are small. More than one hundred and fifty kinds inhabit North America. They devour all sorts of edible substances, animal as well as vegetable, and some even attack living animals.

The largest, except the Muskrat, is the Norway, Brown, or Wharf Rat, originally from Asia, but now exceedingly abundant in Europe and in this country.

Fig. 93. — White-footed Mouse.

The Black Rat, which was introduced into this country from Europe more than three hundred years ago, is nearly as large as the Brown, and was formerly the most common large Rat in stores, houses, barns, and other buildings, but has now nearly disappeared before its more powerful rival, the Brown Rat, which pursues, captures and devours it. If a rat gets wounded, his companions, instead of aiding him, fall upon and devour him. The Roof Rat, of the Southern States, originally from Egypt, where it lives in the thatched roofs of the houses; the House Mouse, originally from Asia, but now found in all countries; the Harvest

Mouse, the White-footed Mouse, the Field Mouse, and the Jumping Mouse, are other kinds which are found in the United States, but which cannot be described here for want of room. For further description, see Tenney's Manual of Zoölogy. The Jumping Mouse, however, is too interesting to be omitted. It is found over a large part of North America, and is about three inches long to the tail, which, in some instances, is even six inches in length. Its color is yellowish-brown, lined with

Fig. 94. — American Jumping Mouse.

black, the lower parts white. It moves by very long and rapid leaps. It is found in the meadows and grain-fields.

The Muskrat, mentioned above, is very common about ponds, rivers, and brooks in North America. It is a foot long, besides the tail, which is about as long as the body, and the color is dark brown above and rusty brown below. The fur is now sold under the name of river sable. Muskrats build winter houses of mud, sticks, and grass, the entrance being beneath the water, and leading to a dry apartment above.

Porcupines.

Porcupines are distinguished from all other Rodents by their spines, or quills, which are very sharp. The North American Porcupine is about two feet long, brown in color, with long white-tipped hairs, and has the tail and upper parts covered with white spines. It lives in hollow trees and in holes among the rocks, and readily climbs trees. It eats bark, leaves, and green corn. It is often called the Hedgehog, but is a very different animal from the true Hedgehog (p. 60). See Figure 95. The Crested Porcupine, of Southern Europe, has quills nearly a foot long.

Fig. 95. — American Porcupine.

Hares.

Hares are found in nearly all countries. In America there are about thirty kinds. They are timid, and have a habit of stamping with the hind feet when alarmed. The Common Hare, or White Rabbit, about twenty inches long, is brown in summer, and white in winter. It lives in the thick swamps, rarely enters holes when pursued, but depends for safety upon its fleetness. It always follows the same paths. The Gray Rabbit is smaller and does not turn white in winter.

Toothless Mammals, or Edentates.

The Edentates are Sloths, Armadillos, and Ant-eaters. Some of these animals have no teeth, and others are only destitute of front teeth. Many of them have a bony or scaly covering. They live in warm countries.

Armadillos.

The word Armadillo means *clad in armor*, and is given to these animals on account of their bony or horny covering. They live mainly in the warm and

Fig. 96. — Nine-banded Armadillo.

hot parts of South America, dig burrows, and feed upon vegetables, carrion, insects, and worms. The Nine-banded Armadillo is about two feet long, and is found as far north as Texas.

Marsupials.

The Marsupials have a pouch, or sack, beneath the body, in which the young are kept for a time after they are born; even after they are able to walk they resort to the pouch of the mother when danger is near. With the exception of the Opossums of America, all the Marsupials are in Australia and adjacent islands.

Opossums.

Opossums are small animals, the largest being scarcely larger than the Common Cat, and the smallest but little larger than a Mouse. They feed upon birds, bird's eggs, insects, and other small animals. The tail is long and is capable of being twisted around objects, thus aiding in climbing. The Opossum of the United States is about the size of a Cat, the hair whitish with brown tips. It often lies motionless for hours in the warm sunshine. When slightly wounded it has the habit of feigning itself dead, or "playing 'possum," and often escapes from the inexperienced hunter.

Fig. 97. — Opossum.

Kangaroos.

Kangaroos are Marsupials which are remarkable for the great development of their hinder parts,— the hind

Fig. 98. — Kangaroo.

legs and tail being very long and powerful, and the fore legs very short, weak, and but little used in locomotion, which is accomplished by leaps of enormous extent. They live in troops, feed upon vegetation, and are harmless and easily tamed. They vary in size from that of a Rabbit to that of a Deer.

Fig. 99. — Wombat.

Fig. 100. — Skull of Wombat.

The Wombat is a curious Australian animal, three feet long. Its habits are not unlike those of the Woodchuck: it feeds upon grass, and burrows in the ground.

MONOTREMES.

These are animals which vary much from all other Mammals, having their organic structure in some respects much like that of Birds. They belong to Australia and adjacent islands. One of the most interesting

Fig. 101. — Duckbill, or Platypus.

kinds is called the Duckbill, or sometimes Platypus. Its muzzle is flat and appears very much like that of a Duck, its legs are short, its feet webbed, and its body is covered with short, brown fur. It is less than two feet long, lives about ponds and streams, and digs burrows in the banks. Its young are born from eggs which it lays in its burrows. The only other member of this group, the Spiny Ant-eater of New Guinea, lays an egg which it carries in its pouch until hatched.

BIRDS.

Of all animals, perhaps, none are more interesting to both young and old than Birds. Their presence in the fields and hedges, the groves and forests, their beautiful and splendid colors, their sweet songs, and their curious and wonderful habits, charm and delight all.

Birds are egg-laying vertebrates which are covered with feathers, furnished with a bill, and fitted for flight, — their form as well as their structure being adapted for easy and rapid movement through the

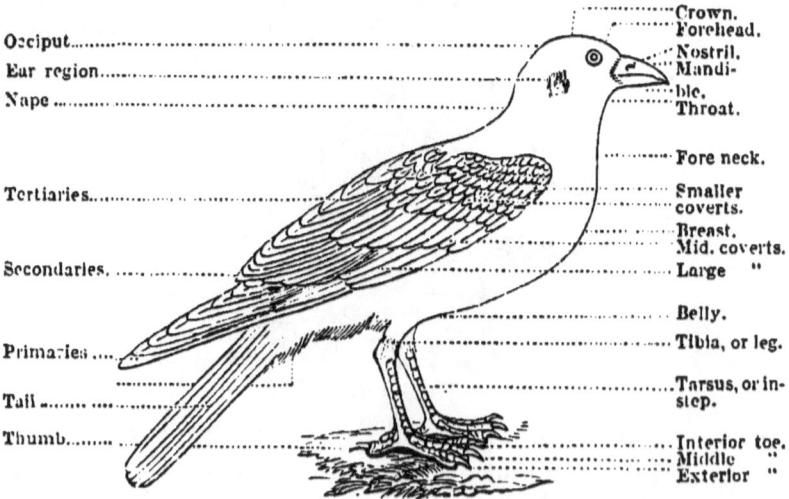

Fig. 102. — Showing the names of some of the principal parts of a Bird.

air; even their bones are hollow, hence very light in proportion to their size. The general form of a Bird, and the names of some of the principal external parts, are shown in Figure 102. The skeleton and the names of its principal parts are shown in Figure 103. It is an interesting fact that the form and the skeleton of a bird suggested the right way in which to build a ship in order to combine strength with swiftness.

Although the body of Birds is covered with feathers, these do not grow from the whole surface, but are arranged in rows and patches, with bare spaces between. Feathers are made up of a hard central portion, or shaft,

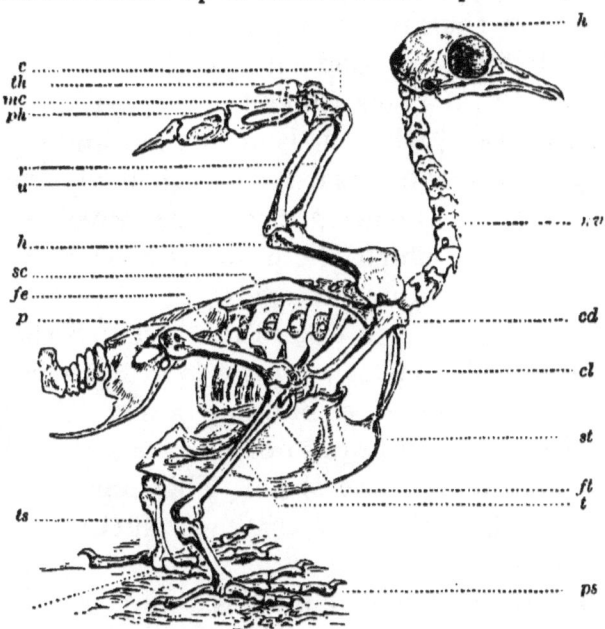

Fig. 103. — Skeleton of a Bird.

h, head; *nv*, neck vertebræ; *c*, wrist; *th*, thumb; *mc*, metacarpus, or hand; *ph*, phalanges, or fingers; *r*, radius; *u*, ulna; *h*, humerus; *sc*, scapula, or shoulder-blade; *cd*, corocoid bone; *cl*, clavicle, or " wish bone "; *st*, breastbone, or sternum; *fe*, femur, or thigh bone; *p*, pelvis; *ft*, fibula and tibia united; *t*, tibia, or leg; *ts*, tarsus, or instep; *ps*, phalanges, or toes.

and vane on each side, the latter being the broad portion which consists of delicate plates that are united by minute barbs along their edges, and thus made firm,— the plates not separating from one another when pressed against the air, as in flying. There are, however, downy feathers on every Bird, or such as do not have the plates united. The plumage of Birds is made waterproof by the oil with which they dress their feathers, and which is obtained from a gland situated on the tail. They shed their feathers twice a year, and in many kinds the winter plumage differs in color from that of the summer. In most Birds the colors of the male are much more brilliant than those of the female.

Birds swallow their food without chewing it, and it is first received into a sack called the crop; then it passes into another sack, where it is moistened and softened; then it passes to the gizzard, where it is digested. In seed-eating Birds, the gizzard contains gravel and other hard substances, which these animals swallow to aid the gizzard in grinding the seeds.

Birds lay eggs and sit upon them to hatch them, and most Birds build nests in which to rear their young, those of the same kind building alike. The young Bird in the egg has a horny point at the end of the bill, with which it breaks the shell. This point is plainly seen on the bill of the newly-hatched chicken; in a few days it falls off.

The number of kinds of Birds is about ten thousand, and there are about seven hundred kinds in North America. Birds of Prey, the Climbers, the Perchers, the Scratchers, the Runners, the Waders, and the Swimmers are the large groups into which Birds are divided.

BIRDS OF PREY, OR RAPTORES.

These are the Vultures, Eagles, Hawks, Falcons, and Owls. Most of them capture birds and other animals for food. They are mostly of large size, and have a strong hooked bill, sharp claws, great spread of wing, and very powerful muscles, and the females are generally larger than the males. They live in pairs, and choose their mates for life.

Fig. 104. — California Vulture.

Vultures.

Vultures have the head nearly naked or thinly covered with feathers, and, unlike the other rapacious Birds, seldom capture prey, but feed upon carrion, which they trace by sight at great distances. They make no nest, but deposit their eggs on the ground or naked rock. There are three or four kinds in the United States. The Condor of the Andes and the Lammergeyer of the Alps are Vultures of the largest kind. The latter attacks lambs, goats, and the chamois. The California Vulture is the largest Bird of Prey in North America, being as large as the largest Turkey; the color is black, the head orange and red. See Figure 104.

Falcons, Hawks, and Eagles.

These Birds have the head clothed with feathers, and their talons are very sharp. Their flight is rapid, and they attack their prey with great ferocity, capturing chickens, ducks, grouse, quails, hares, rabbits, squirrels, and other small animals. The species are numerous, about seventy kinds of Eagles being known, and more than thirty kinds of Falcons and Hawks inhabiting North America. The true Falcons have a distinct tooth in the upper mandible, as seen in Figure 105.

Fig. 105. — American Peregrine Falcon, or Duck Hawk.

BIRDS OF PREY. 77

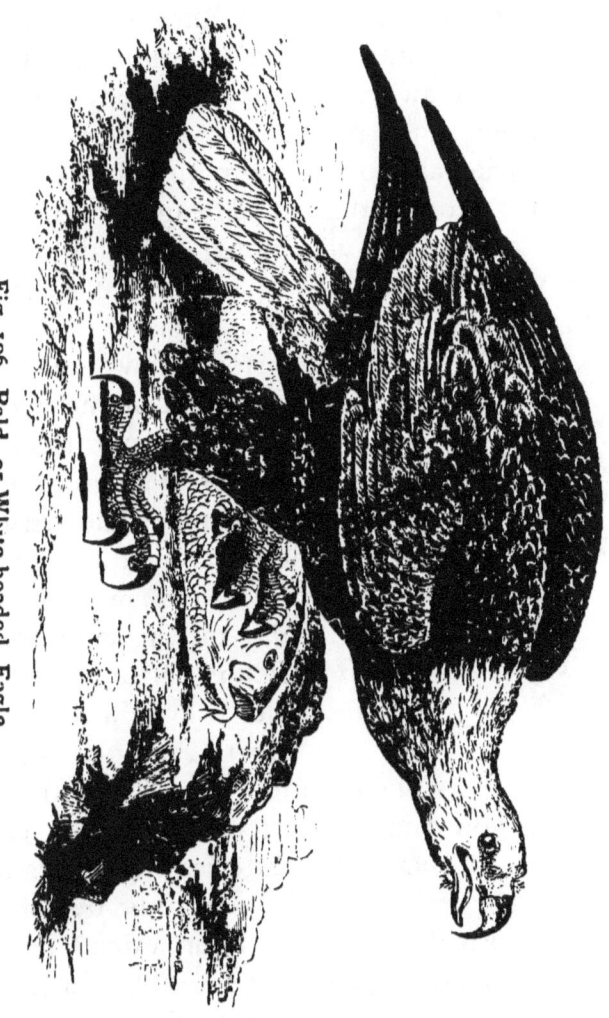

Fig. 106. Bald, or White-headed, Eagle.

The Peregrine Falcon, or Duck Hawk, of North America, pursues its prey with almost inconceivable velocity through all its turnings and windings, and when within a few feet, protrudes its talons, grasps the prize, and bears it away to some secluded place and devours it. Sometimes it sweeps over the water and catches up ducks and other swimming birds. This Falcon is about a foot and a half in length. The Peregrine Falcon of Europe, very much like this species, was formerly much used in falconry, a fashionable sport of kings, nobles, and fair ladies.

Fig. 107—Sparrow Hawk.

The Sparrow Hawk, of America, is the smallest of the Hawks, being but little larger than the common Robin. It preys upon small birds, mice and insects. It becomes attached to a particular locality, and may be seen day after day on the same tree or stump watching for prey.

The Bald, or White-headed, Eagle, of North America, is found along the seacoasts, lakes and rivers, and usually makes its nest on some tall tree. Although called bald, its head is clothed with white feathers. Its principal food is fish, which it obtains mainly by robbing the Osprey, or Fishhawk. Seated on a dead limb of a large tree that commands a view of the waters, it watches the Fishhawk as he descends and plunges into the deep, and, as he emerges with his prey and rises into the air, the Eagle gives chase; each moves with its utmost speed, but the Eagle rapidly gains, and as it is about to reach the Hawk, the latter drops the

Fish; the Eagle sweeps downward, snatches it before it reaches the water, and bears it away to the woods.

Owls.

Owls are Birds of Prey which, in most cases, are active by night, and rest during the day. Their large head, large staring eyes, and the tufts of feathers resembling ears, which many of them have, give to the

Fig. 108. — Great Horned Owl.

face a strange, catlike expression. Their plumage is soft and loose, and their flight is almost noiseless. They prey upon birds, hares, squirrels, mice, and insects. There are about forty kinds of Owls in America, greatly varying in size. The Great Horned Owl has large ear-tufts like horns; the Screech Owl is small, and is noted for its tremulous, doleful notes; the Long-eared Owl has very long ear-tufts, and its cry is prolonged and plaintive, consisting of two or three notes repeated at intervals; the Gray Owls are very large; the Saw-whet Owl is small, and its notes sound like

Fig. 109. — Snowy Owl.

the noise made in filing a saw; the Burrowing Owls are very small, and live in the burrows of the Prairie Dog. The Snowy Owl is large, and hunts in the daytime as well as at twilight; it lives in the cold regions, and is seen in the United States only in winter.

CLIMBERS, OR SCANSORES.

Climbers have two toes in front and two behind. Parrots, Cuckoos, and Woodpeckers are the chief kinds.

CLIMBERS. 81

Parrots.

Parrots have a stout, thick bill, hooked at the tip. Many of them are adorned with the most gorgeous colored plumage; and this, together with the ease with which many of them are trained to speak, has made them objects of great interest. They live in the warm regions.

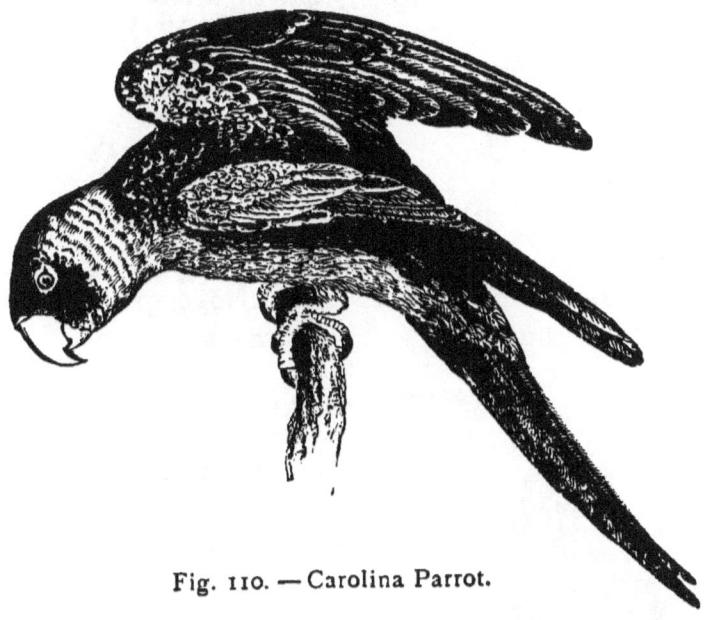

Fig. 110. — Carolina Parrot.

The Carolina Parrot of the Southern States, about as large as a Dove, is our only species.

Cuckoos.

The Cuckoos of the United States are about a foot long, with the upper parts of a metallic olive-green color, and the under parts white. They are shy, concealing themselves in the thick foliage of trees, where they sit for hours uttering their unpleasant notes, which sound like *cow-cow*, eight or ten times repeated.

Fig. 111.—Cuckoo.

They feed upon insects, and also eggs, which they steal from the nests of other birds.

Woodpeckers.

Fig. 112.—Red-headed Woodpecker.

These Birds have a straight, sharp bill, with which they cut into bark or wood in search of insects. The tongue is very long and capable of being greatly extended, and is armed towards the tip with barbs. With this they secure the insects. Twenty or thirty kinds are found in North America, varying in size from the Sparrow to the Crow. They build their nests in holes, made with their bills in the trunks or branches of trees.

The Ivory-billed Woodpecker, of the Southern States, is the largest, and has the body black with white upon the wings and neck, the crest scarlet, and the bill ivory white. The Black Woodcock, of the Northern States, is smaller, greenish-black in color, with a scarlet crest. The Hairy and the Downy Woodpecker, or Sapsuckers, are small, and black and white. The Red-headed Woodpecker has the head and neck, crimson; the back, primaries and tail, black; the rump and a band on the wings, white. The Golden-winged Woodpecker is larger

Fig. 113. — Golden-winged Woodpecker.

than a Robin, and is one of our most beautiful Birds. On the first sunny days of spring the Woodpeckers of this species appear on the top of decayed trees, and as they hop about, striking with their bills here and there, make the woods resound with their loud, clear notes. Soon they pair, and both male and female begin to make a hole in a tree for the nest. The female lays from four to six white eggs twice in a season.

PERCHERS, OR INSESSORES.

These make up a large part of the most common Birds, as Humming Birds, Nighthawks, Kingfishers, Flycatchers, Thrushes, Warblers, Creepers, Titmice, Sparrows, Grosbeaks, Larks, Blackbirds, Jays, Crows, etc.

Humming Birds.

These are Birds of the smallest size and of the most gorgeous plumage to be found in the feathered race. The beauty of their colors defies description; and from their brilliancy they are often called "flying gems."

Figs. 114 and 115. — Ruby-throated Humming Bird and Nest.

There are about four hundred kinds, and they all belong to the continent and islands of America, and are most numerous in the warm regions. Their feet are very small, their wings long, and their power of flight very great; they can balance themselves in the air, or beside a flower, with perfect ease. Their food consists of insects and the honey of flowers. Their nests are usually made of cotton, thistle-down, delicate fibers, and other soft materials, woven into a cup-shaped cradle, and placed on a branch of a tree not many feet from the ground; and the outside is covered with lichens to make the nest appear like a natural growth. The eggs are pure white. The Ruby-throated Humming Bird is common throughout the United States.

Whippoorwills and Nighthawks.

The Chuck-will's Widow, whose curious notes are heard in the evening and in the early morning in the

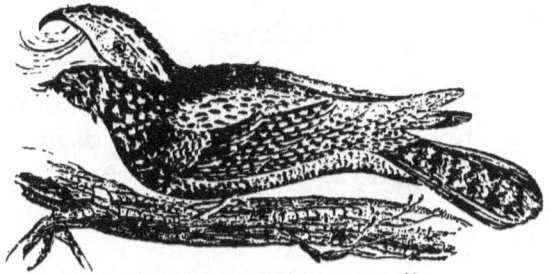

Fig. 116. — Whippoorwill.

Southern States, and the Whippoorwill and Nighthawk of the United States generally, are closely related to

Fig. 117. — Nighthawk.

each other. The last two are each about ten inches long, and dark, marked with white. The Chuck-will's Widow gets its name from its notes, which sound like *chuck-will's-widow*, and the Whippoorwill from a resemblance of its notes to the syllables *whip-poor-will*, uttered in the evening and at dawn. They make no nest, but lay their eggs on the ground, or a flat rock.

Kingfishers.

These Birds feed upon fish, and make their nests in holes which they dig in the banks of ponds and streams. They have a long, straight bill, and short legs. The Belted Kingfisher, of North America, is nearly as large

Fig. 118.— Belted Kingfisher.

as a small Dove, the head crested, the color blue above and white below, with a blue belt. Sitting on a branch or decayed tree near the water, it watches intently for fish; and at the proper moment it plunges into the water, seizes its victim, flies to a tree, swallows the fish, and is immediately on the lookout for another.

Flycatchers.

There are about thirty kinds of these Birds in North America. The bill is broad and bent down at the tip, and the sides of the mouth have stiff bristles. The Kingbird, Pewees, and Great-crested Flycatcher are some of the most common and best known species. The Kingbird is somewhat smaller than a Robin, and

Fig. 119. — Kingbird.

is dark above and white below, with a hidden crest of orange, vermilion and white. It is common in open fields and orchards, where it is seen perched upon a stake, tall weed, or low tree, watching for insects, upon which it darts with sure aim. It is very courageous, eagerly attacking crows, hawks, and other large birds.

The Pewee, or Phœbe Bird, is smaller than the King-bird, and its color is dark above and yellowish below.

It lingers around bridges, old buildings, and caves. Here in some secure spot it builds its nest of mud, grass, and moss, with a soft lining within for the eggs, which are pure white with reddish spots near the larger end. The Wood Pewee is rather smaller than the Phœbe, and is found in the quiet retreats of the forest.

Thrushes, Bluebirds, and Robins.

The Wood Thrush, Hermit Thrush, Wilson's Thrush, Robin, Robin Redbreast, etc., come under this head.

The Wood Thrush is smaller than a Robin, brownish above, white below, marked with triangular black spots.

Fig. 121. — Ruby-crowned Wren.

Fig. 120 — Wood Thrush.

It is found in groves and woods, and its sweet singing has made it celebrated among all lovers of birds. Its nest and eggs much resemble those of the Robin.

The Hermit Thrush is smaller than the Wood Thrush, which it somewhat resembles, but it is rather darker above, its breast is yellowish-white, and the dark spots beneath are less distinct. Its soft, liquid, plaintive notes excel in sweetness those of any other American Bird. It is heard in shady glens and deep woods.

The American Robin is one of the most common of

the Thrushes, and its song in the early morning and at the close of the day is one of the pleasantest sounds that come from our groves and orchards.

The Robin Redbreast, of Europe, is about half as large as our Robin, of a brown color, with a red breast. It is nearly related to our Bluebird. It loves to be near man, and often enters his dwelling. It is easily tamed, and is a great favorite. In severe weather it comes into the house, and, selecting a perch, warbles its song when the day is clear or when the fire burns brightly.

The American Bluebird is sky-blue above; the breast is chestnut-colored. Its nest is usually made in a hollow tree or post, and its eggs are from four to six, pale blue. It is a loving, gentle Bird, and its soft warble is very pleasing. The Ruby-crowned Wren, which is now placed near the Thrushes, is scarcely more than four inches long, and is known by a patch of scarlet feathers on the crown. Its song is clear and sweet. The Water Ouzel, of the Rocky Mountains, is smaller

Fig. 122.— Ouzel.   Fig. 123.— Nightingale.

than the Robin, and of a dark lead-color. This curious Thrush frequents mountain streams, into which it walks or dives, and moves about beneath the water in search of insects and other small animals upon which it feeds.

Warblers.

Warblers are among the smallest, most beautiful, and interesting of singing Birds. Many kinds are generally found in the same locality, and may be seen gliding among the thick foliage, busily engaged in catching minute insects which hide beneath the leaves and in the buds and blossoms, and which often escape the sight of other and larger Birds. Some of the Warblers are the sweetest of songsters, as the celebrated Nightingale of Europe, shown in Figure 123. More than fifty kinds are found in the United States; their very names are beautiful and suggestive. Some of the more common are the Maryland Yellowthroat, the

Fig. 124. — Maryland Yellow-throat.

Fig. 125. — Blackburnian Warbler.

Blue-winged Yellow Warbler, the Golden-winged Warbler, the Orange-crowned Warbler, the Golden-crowned Warbler or Thrush, the Black-throated Green Warbler, the Yellow-rump Warbler, the Bay-breasted Warbler, the Chestnut-sided Warbler, the Blue Warbler, the Blackpoll Warbler, the Yellow Warbler, the Black and Yellow Warbler, the Yellow Redpoll, the Yellow-throated Warbler, the Blackburnian Warbler, etc.

Swallows.

These beautiful Birds have long wings, short legs, and short, wide bill, and they spend much of their time upon the wing, skimming over fields and ponds, catching small insects, which constitute their food. One kind builds its nest upon the rafters in the barn, and is called the Barn Swallow; another builds its nest under the eaves, and is called the Eave Swallow; another under cliffs, and is called the Cliff Swallow; another digs a hole in a sandbank for its nest, and is called the Bank Swallow; and the Purple Martin comes and makes its nest in the martin houses placed for it near our dwellings. Some persons suppose that these birds, requiring air and sunshine as much as we do, spend the winter in the mud at the bottom of ponds.

Shrikes and Vireos.

The Shrike, or Butcher Bird, is about as large as a Robin, of a bluish color, with black wings and tail. Although belonging to the song-birds, it is a hawk in

Fig. 126. — Shrike, or Butcher Bird.

its disposition, preying upon sparrows, warblers, and other small birds, as well as upon insects. It often imitates the cries of other birds, perhaps to call them from the trees and bushes, that it may get a chance to seize one of their number. It is called Butcher Bird from its habit of impaling or hanging up its prey upon thorns and other sharp points, as a butcher hangs meats upon hooks in his stall. It builds a large nest of twigs, grass, and moss, in the forks of a tree.

Fig. 127. — Warbling Vireo.

The Vireos are much smaller than the Shrike, and mostly olive-green above and light below. The Red-eyed Vireo has the iris of the eye red. Its loud, clear notes are heard in the tree-tops from spring till late in autumn. The White-eyed and the Warbling Vireo are small, and their notes are very pleasant.

Mocking Birds, etc.

These Birds are closely related to the Thrushes, and are very sweet singers. The Mocking Bird of the Southern States is about the size of the Robin, with a

Fig. 128. — Mocking Bird.

very long tail, and the color is ashy. It sings with great sweetness, and readily imitates the songs of all the birds which it hears. It is a very common pet in cages.

The Catbird of the Northern States is smaller than the Robin, and of a dark color, and in spring and the early part of summer its song is very mellow and sweet. Like its relative, it easily imitates the notes of other birds, and may be properly called the Mocking Bird of the North. President Hill, of Harvard College, states that, having whistled a strain of Yankee Doodle two or three times in the presence of this bird, it imitated him perfectly. In the latter part of summer its notes are harsh and disagreeable, sounding like a cat's yawl.

Wrens are small Birds about the size of the Warblers. The Carolina Wren is one of the largest. It is reddish-brown. The House Wren delights in being near the habitations of man, and often makes its nest in a hole in the timbers or walls. The Winter Wren is one of the smallest, and of a brownish color. It is very active, and may be seen in twenty attitudes in a minute.

Fig. 129. — Winter Wren.

Creepers, Nuthatches, and Chickadees.

Creepers and Nuthatches are very small Birds, which may be seen in North America at all seasons of the year, running along the trunks and branches of trees, and looking, at a little distance, much like little Woodpeckers. The American Creeper is light brown, with lighter streaks. The White-bellied Nuthatch is blue, with the under parts white, and the top of the head and neck black. The Red-bellied Nuthatch is a smaller species, and has the under parts red. Both kinds attach their feet to the bark, and creep with their heads downward. The Chickadee is one of our smallest birds, and sings its simple *chickadee-dee-dee* in winter as well as in

Fig. 131. — Chickadee, or Titmouse.

Fig. 132. — White-bellied Nuthatch.

Fig. 130. — American Creeper.

summer, and in all sorts of weather. It is ashy above, whitish below, the top of the head and throat black.

Skylarks.

The Skylark, or Shore Lark is the only Bird of its family in North America. It is smaller than the Robin, and sings sweetly while on the wing, but its song is

Fig. 133. — American Skylark.

short. The Skylark of Europe is almost as celebrated for its song as the Nightingale. It often rises vertically to a great height, and when rising or falling it sings its varied and powerful song.

Finches, Crossbills, Buntings, Sparrows, and Grosbeaks.

The Purple Finch is about as large as the Bluebird, and of a beautiful crimson color; the female brown above and white below streaked with brown. The nest is built in a tree close to the ground, and the eggs are four, of a rich green color. The Yellowbird, or Amer-

Fig. 134. — Purple Finch.   Fig. 135. — White-winged Crossbill.

ican Goldfinch, is of a beautiful yellow, the crown and wings black, tail and wings marked with white. The nest is very handsome, made of lichens, and fastened to a twig; eggs white, with a bluish tinge, and spotted with brown at the larger end.

Crossbills have the points of the bill much curved and crossing each other. By means of this curious instrument they can open the cones of pine and spruce with great facility, and thus secure the seeds, upon which they feed. Crossbills are about as large as the

Bluebird; and there are two species in North America, — the Red Crossbill and the White-winged Crossbill, the latter having white bands upon the wings.

Fig. 136. — Song Sparrow.

Sparrows are plain-colored birds, generally dull brown, variously striped and marked, and are the most common in open fields, orchards, and about low bushes. There are many kinds in North America, all of which are small, the largest scarcely equalling the common Bluebird in size. Some of the principal kinds are the Bay-winged Bunting, the Yellow-winged Sparrow, the White-crowned Sparrow, the White-throated Sparrow, the Black Snowbird, the Tree Sparrow, the Chipping Sparrow, the Song Sparrow, the Swamp Sparrow, the Fox-colored Sparrow, etc.

The Grosbeaks have the bill very large, and hence their name, which means *great beak*. The Rose-breasted Grosbeak is one of the most beautiful of the North American birds. It is smaller than a Robin, and the color is black and white, the breast a rich carmine. The female has no black or carmine. The song is loud, clear, and sweet.

The Ground Robin, Towhee Bunting, or Chewink, is about two thirds as large as a Robin, the color black and white. The fe-

Fig. 137. — Rose-breasted Grosbeak.

male is brown and white. It is seen almost everywhere, in low bushes, in fields, or by the wayside, and is easily found by its sweet *chewink*, uttered every few moments. Often near the close of day in spring, it mounts the top of a small tree, and sings with charming sweetness.

Fig. 138. — Chewink.

It makes its nest upon the ground, laying from four to six eggs of a light color with dark spots.

Blackbirds, Larks, etc.

The Bobolink, Cowbird, Blackbirds, Larks, and Orioles belong to one family. The Bobolink is somewhat larger than a Bluebird, of a black and cream color, the female yellowish brown. Its jingling song, uttered from a low tree, or bush, or tall weed, or upon the wing, is familiar to all who live in the country. Late in the summer Bobolinks fly southward, and are seen in immense flocks in grain fields and along the margins of creeks and rivers, where the tops of the reeds are bent with ripe seeds. Thousands are shot by the hunters and sold in the markets, where they are called Reedbirds.

Fig. 139. — Bobolink, or Reedbird.

The Cowbird is larger than the Bobolink, and is the most singular bird in North America. For some reason which is not understood it never makes a nest, but, like the European Cuckoo, stealthily lays its eggs, only one in a place, in the nests of Warblers, Flycatchers, Bluebirds, Sparrows, and the Golden-crowned Thrush. The egg is grayish blue marked with brown dots and short streaks. And it is a curious fact that this egg hatches before the eggs of the bird in whose nest it is laid. As soon as the young Cowbird is hatched, the foster parents leave their own eggs to get food for it, and hence the young in their eggs die, and the eggs are soon thrown from the nest. Then the young Cowbird receives the whole attention of those that have been compelled to adopt it, and they feed it till long after it can fly, and until it is larger than the foster parents themselves. The head and neck of the Cowbird is of a chocolate color, the rest of the body lustrous black; the female is light brown.

The Red-winged Blackbird is nearly as large as the Robin, shining black, with the shoulder and a part of the wing bright crimson. The female is of a dusky color. It is common about ponds and marshes, and builds its nest in low bushes or tufts of sedges.

Fig. 140. — Meadow Lark.

The Meadow Lark is rather larger than the Robin; the upper parts brown and brownish white, the under parts yellow, with a black crescent upon the breast. The nest is built at the foot of a tuft of grass, and is covered over, except the entrance.

The Baltimore Oriole, or Golden Robin, is as large as a Sparrow, the color black and orange-red, and is one of the most beautiful Birds in the United States. Its song is loud, full, and mellow. Its hanging nest, often made from the silkweed, is woven to the outer-drooping twigs of trees.

Crows, Ravens, Jays, and Magpies.

These are rather large Birds. The Raven is the largest. It is but seldom seen east of the Mississippi. The Crow is well known, and farmers regard it as their enemy, because it pulls up the young corn; but it does much more good than harm, by destroying a great number of grubs, which would injure the crops. The Blue Jay is a bird of wonderful beauty, but its notes are harsh, it eats the eggs of other birds, and even destroys young birds, swallowing them greedily. The Magpie is about as large as a Dove, black and white, and the tail is very long. There are two kinds in North America, and one in Europe. It is a noisy bird, and it can be taught to speak.

Fig. 141. — Magpie.

SCRATCHERS, OR RASORES.

Doves, Wild Pigeons, Turkeys, Hens, Grouse, Pheasants, and Quails are the principal Rasores. Most of them live mainly upon the ground, and all feed upon seeds, grain, nuts, and berries. The Rasores furnish man with some of his choicest food. Excepting the Doves and Pigeons, they can run as soon as hatched.

Pigeons.

The Wild Pigeon of North America is about as large as a Dove, and has a very long tail. The color above is

Fig. 142. — Wild Pigeon.

blue, under parts reddish, and the neck glossy, golden-violet. It flies very rapidly, and millions used to be seen, moving together, darkening the air like a cloud. On alighting, they would fill forests, and even break down large trees by their weight. They are now found only in small numbers.

Grouse.

The Prairie Chicken, Ruffed Grouse, Ptarmigans, etc., come under this head.

The Prairie Chicken is about as large as a common Hen, and the male has an air sack on each side of the neck by which it is able to produce a loud booming sound. The Ruffed Grouse, or Partridge, of the United States is rather smaller than the common Hen, and has a beautifully barred and spotted plumage. This Bird prefers open woods and the borders of forests, and in

Fig. 143. — Ruffed Grouse.

winter thickets of evergreens. When disturbed it takes wing with a loud whir. In the spring the male, while standing upon an old log, makes a loud sound with his wings, which is called drumming. The female makes her nest of leaves upon the ground, and lays a dozen or more dingy-white eggs.

Fig. 144. — Quail.

Quails.

These Birds are much smaller than the Grouse. The Quail has a body about as large as a Pigeon, and its

Fig. 145. — Mountain Quail.

color is reddish-brown. In the south it is called the Partridge. Its notes are a sort of whistle. The nest is built near a tuft of grass, and the eggs are from ten to eighteen, pure white. The Mountain Quail is found in Oregon and California.

Runners, or Cursores.

These are the Ostriches and their relations. They are very large Birds with long legs and rudimentary wings. The Camel Bird, or great Ostrich of the deserts of Africa and Asia, is about eight feet high, and has only two toes to each foot. The Rhea is a three-toed Ostrich of South America. The Cassowaries are three-toed Ostriches which inhabit the Indian Archipelago and Australia. The Apteryx is a small ostrich-like bird of New Zealand. Gigantic birds of this group, now extinct, lived in Madagascar and New Zealand.

Waders, or Grallatores.

The Waders have a long bill, long neck, and long legs. They are the Cranes, Herons, Ibises, Plovers, Turnstones, Stilts, Woodcocks, Snipes, Yellowlegs, Godwits, Curlews, Rails, and Gallinules. They live mainly upon marshes or shores, are adapted by their long legs for wading, and feed upon worms, shellfish, etc. Figures 146–158 show some common kinds.

Herons.

The Great Blue Heron, of North America, frequents ponds and creeks, where it may be seen standing for hours, upon a rock or stump, watching for fish. When wounded it is dangerous to approach it, as it strikes with its bill, and generally aims at the eye. This Heron is four feet long. It builds its nest on a large tree, in a dense swamp.

Fig. 146. — Great Blue Heron.

WADERS.

The Bittern, or Stake-driver, and the Night Heron

Fig. 147. — Bittern, or Stake-driver.

with its long, white plumes, are much smaller species.

Ibises.

The Wood Ibis is nearly as large as the Great Blue

Fig. 148. — Wood Ibis.

Fig. 149. — Plover.

Fig. 150. — Turnstone.

Fig. 151. — Yellowlegs.

Fig. 152. — American Woodcock.

Fig. 153. — Wilson's Snipe.

Fig. 154. — Stilt.

WADERS.

Fig. 155 — Godwit.

Fig. 156 — Curlew.

Fig. 157. — Rail.

Fig. 158. — Gallinule.

Heron, and lives in the swamps of the Southern States. In order to obtain food, it moves about in the shallow waters until these become muddy, when the fishes rise to the surface, and are struck and killed by its bill.

Swimmers, or Natatores.

These Birds are fitted to live in and about the water. Their feet are webbed, and the plumage is thick and made waterproof by the oil with which they dress it. They swim easily, and most of them are expert divers. Swans, Geese, Ducks, Pelicans, Petrels, Gulls, Divers, Auks, and their relatives, belong in this group.

Swans, Geese, and Ducks.

The Swans have the neck very long, and they are much larger than the largest Goose. There are two species in North America, — the American Swan and the Trumpeter, both pure white.

The Wild Goose is larger than the Common Goose, of a brownish color, with black head, neck, bill, feet, and tail. Wild Geese are seen in early spring in large flocks, moving northward, where they rear their young, returning south in autumn. The peculiar noise made by a flock as they pass over is familiar to all. They are sometimes tamed, but often manifest a desire to join the migrating flocks. A wild goose was kept all winter with a flock of common geese. The following spring it joined a party of its own kind which was passing over. The next autumn, as a flock of wild geese was returning southward, three of the number separated from the others and alighted in the poultry yard.

They proved to be the long-lost goose, and two of her young.

The Mallard, or Greenhead, is about two feet long, and has the plumage of the head bright green; there is a white ring around the neck, and the general color of the body is brownish. This is the parent of the Domestic Duck.

The Wood Duck is smaller than the Greenhead, and its plumage excels in beauty that of all other Ducks.

Fig. 159. — Wood Duck.

It builds its nest in a hollow tree or limb; and if the nest is over water, the young, as soon as hatched, drop into it; if not, they fall to the ground, and are led or carried to the water by the parent.

The Canvasback is about the size of the Wood Duck, with a chestnut-colored head, and the other parts white and black.

The Eider Duck is one of the largest of the Ducks; colors, black and white. It lives in the cold north

Fig. 160. — Canvasback.

regions. Eider down comes from the nests; the birds pluck it from their breasts to place around the eggs.

Albatrosses and Petrels.

The Albatrosses are the largest of web-footed Birds.

Fig. 161. — Sooty Albatross.

The Petrels, in many cases are very small. Both live on the ocean, but come on shore to rear their young.

The Stormy Petrels, or "Mother Carey's Chickens," are the smallest of web-footed birds; but they are able to fly about during the most terrific storms. While flying close to the water they extend their legs, and thus appear to walk upon its surface. The word *Petrel* means *little Peter*.

Fig. 162. — Stormy Petrel.

Gulls and Terns.

The Gulls and Terns have long and pointed wings, and are common upon the shores of all countries, and

Fig. 163. — Tern.

also on the larger rivers and lakes. They swim well, but do not dive. The Gulls are generally light-colored, and they vary in size from that of a Dove to that of a

Goose. The Terns have the tail very long and forked. They are generally light below, black and bluish above, and of the size of a Dove, but some are no larger than a Robin. They feed upon small marine animals.

Divers.

The Great Northern Diver, or Loon, is almost as

Fig. 164. — Great Northern Diver, or Loon.

Fig. 165. — Crested Grebe.

large as a Goose, black above, beautifully spotted with white, and white below. It is exceedingly keen-sighted and wary, and it dives so quickly that, seeing the flash of the gun, it is often under water before the shot reaches it.

Grebes are Divers which are smaller than the Loon, and in the spring have the head ornamented with tufts of feathers. When alarmed, they remain beneath the surface of the water, exposing only the bill.

Auks, Puffins, and Penguins.

These belong to the cold regions, and the Penguins to the southern hemisphere. The Great Penguin of Patagonia is larger than a Goose. Its wings are so

Fig. 166. — Patagonian Penguin.

small that it cannot fly, and it stays in the water most of the time. It is extremely active in the water, often moving in leaps like those of a porpoise. Its eggs are laid in a shallow depression in the sand. Immense numbers of Penguins roost together, occupying with their nests the whole ground, except narrow walks leading to the water.

Fig. 167. — Puffin.

The extinct Great Auk of the Arctic regions was as large as the Penguin; its bones and eggs are rarities in museums. It has not been seen alive since 1844, when the last two were killed near Iceland. Other kinds are much smaller; those called Puffins are not larger than a Dove. The Puffin makes its nest in a burrow and lays but one egg in a season.

REPTILES.

Reptiles are Vertebrates which have cold blood, and are covered with hard plates, called scales, and which lay eggs. In most Reptiles the eggs are not brooded by the parent, and the young, as soon as hatched, look just like the parents, only smaller. Reptiles are such as Turtles or Tortoises, Lizards, and Serpents or Snakes.

TURTLES, OR TORTOISES.

Turtles, or Tortoises, are Reptiles which have a shell into which they can withdraw their head, legs, and tail. Some of them live wholly upon land, like the Gophers in the Southern States which dig burrows that are dangerous pitfalls for horsemen, and the Box Turtles which live in the woods and can shut their shell so tightly as to entirely hide their extremities, as seen in Figure 170. Others, like the Painted Turtle with its colors of black, yellow, and red, the Wood Tortoise with its beautifully carved scales, the Speckled Tortoise with its black shell ornamented with orange-colored dots, and the Snapping Turtle, live in fresh-water ponds and streams, coming at times upon the land. Others, like the Salt-water Terrapin, so much prized for food, live in salt-water creeks. Others, like the Hawkbill Turtle, the Green Turtles, and the Soft-Shelled Sphargis, live in the ocean, and only come on shore to lay their eggs. The land and fresh-water turtles of North America have the shell from four to six or eight inches long, except the Gophers and Snappers, which are much larger, having the shell a foot and a half or more in length; in some cases, the Snapping Turtle is four feet long from the nose to the tip of the tail. This turtle has the head and neck very large, and the jaws strongly hooked; it is exceedingly powerful, and very voracious, devouring smaller reptiles, fishes, young ducks, and other small animals. When molested it raises itself on its legs, opens its mouth wide, and, throwing the body forward, snaps its jaws upon its enemy with fearful power. See Figure 171.

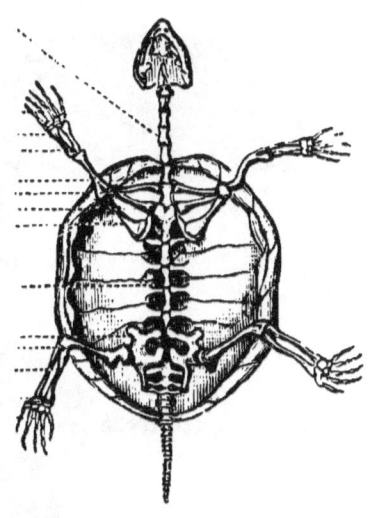

Fig. 168.—Skeleton of a Turtle.

The Hawkbill Turtle, Figure 172, lives in the warm parts of the Atlantic Ocean, and weighs about two hundred pounds; its scales furnish the material for beautiful and costly tortoise-shell ornaments.

The Green Turtles weigh two or three hundred

pounds, or more, and are caught at night when they come on shore to lay their eggs.

Fig. 169. — Wood Tortoise.

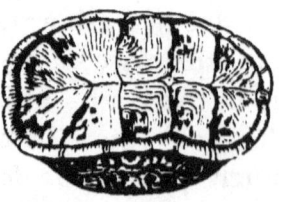

Fig. 170. — Box Turtle, shut up and on its back.

Fig. 171. — Snapping Turtle.

Fig. 172. — Hawkbill Turtle.

The Sphargis, or Soft-shelled Sea Turtle, lives in the tropical regions of all oceans and has been found even in the Mediterranean Sea. It is the largest of all the Turtles, sometimes weighing fifteen hundred pounds.

It is covered with a thick leather-like skin, instead of a hard shell, above the bony case.

CROCODILES AND ALLIGATORS.

These Reptiles have a long body, long tail, teeth set in separate sockets, and a four-chambered heart. The Crocodiles of the Old World have a narrower jaw than the Alligator, and their teeth differ in size. Crocodiles thirty feet long, live in the river Nile. The Alligators, of

Fig. 173. — Alligator.

the Southern States, are five, ten, or fifteen feet long, and have a head shaped something like that of a pickerel. They are numerous in sluggish streams, and devour small animals which come in their way.

LIZARDS.

These animals are small, have a long body and long tail, and are covered only with horny scales.

The Six-lined Lizard, of the Southern States, is only nine or ten inches long, with six yellow lines along its sides and back. It is harmless, runs rapidly, and feeds upon insects. The Green Lizard, of the Southern States, is a smaller species which is common about gardens and buildings, often entering houses, and moving over the furniture, up and down the walls and

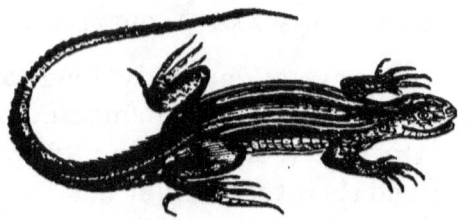

Fig. 174. — Six-lined Lizard.

window panes, and along the ceilings, in its search for flies, upon which it likes to feed.

The Horned Toads are Lizards found in the southern and western parts of North America. The head is armed with spines, the body covered with tubercles. The Horned Toad, of Texas, is less than five inches long, and lively in movement, though sluggish in a cage.

Fig. 175. — Horned Toad.

SNAKES, OR SERPENTS.

Serpents are Reptiles which are exceedingly long in proportion to their size, and which have no feet, yet they glide over the ground with very great speed. They move by the bending of their bodies, aided by the scales which cover their under surface. These are broad and flat and point backward so as to catch on the ground and thus aid in locomotion. Their mouth, throat, and body are capable of being greatly distended, and hence they are able to swallow animals whose bodies are much greater in diameter than their own. They do not masticate their food, and their teeth are suited only for seizing, killing, and retaining prey. The tongue is long, and capable of being run out much beyond the mouth, and it can be concealed within a sheath at its roots. They shed their skins every year, and most of them lay eggs from which the young are hatched. There are more than a thousand kinds of Snake, and more than a hundred kinds in North America. Some of the largest in the

Fig. 176. — Black Snake.

Tropical regions, as the Boas and Anacondas of South America, and the Pythons of Africa and India, are twenty or thirty feet long, and are able to swallow dogs, or even small deer, after they have crushed them in their powerful folds.

The Black Snake and the Striped Snakes are the most common kinds in North America. The former is from three to five feet long, and lustrous black. It runs very fast, and climbs trees and bushes to find bird's nests and devour the young. It is harmless to man.

The Rattlesnake of North America, is found on rocky hills and mountains, and its bite is often fatal to men and animals. It has two very sharp fangs in the upper jaw. These are hollow or grooved, and connected with a bag of poison, so that when the snake strikes them into an animal, the poison is forced into the wound.

BATRACHIANS.

These are Vertebrates which have no scales, and which lay their eggs in the water; the young re-

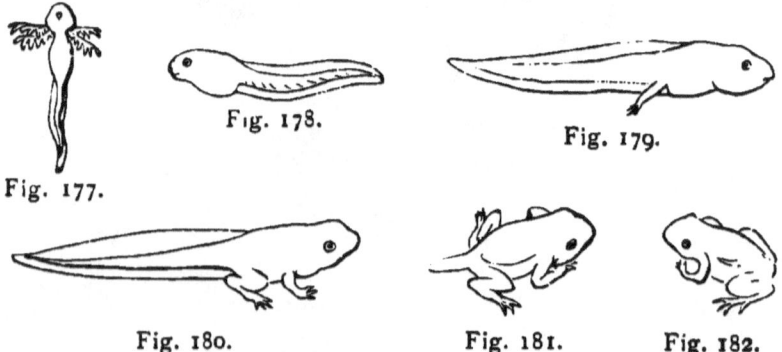

Figs. 177-182.—Changes in the form of a Frog from the time of hatching.

semble Fishes more than they do their parents, and breathe by means of gills, like Fishes; but the adults breathe by lungs. For example, the young frog or tadpole, when first hatched, appears as in Figure 177, with the gills in tufts on the two sides of the neck; later, it appears as in Figure 178, where the gills are concealed; later, it appears as in Figure 179, where it has hind legs; later, as in Figure 180, with four legs; later still, as in Figure 181, where the tail has mostly disappeared; and later still, it becomes a perfect frog.

FROGS AND TOADS.

These have the body short and thick; the tongue is long and fixed to the fore part of the jaw, and its tip is turned backward into the mouth, from which it can be darted forth quicker than a glance of the eye; and it is by means of the tongue that Frogs and Toads snap up insects and worms, which form their principal food. The Bullfrog is our largest kind, and is well known by its croakings, which may be heard a mile. The Green

Fig. 183. — Leopard Frog.

Frog, Leopard Frog, and Pickerel Frog, are found about ponds and streams. The Wood Frog lives on land, and goes to the water only in spring to lay its eggs. The Tree Frogs, or Tree Toads, have toes that enable them to move along the trunks, branches, and leaves of trees. Here they live, except when they go into the water to lay their eggs. One of the tiny Tree Frogs, named Pickering's Hylodes, makes the high piping note, which in spring is heard in New England and in the Middle States throughout the night. It is found upon plants near to stagnant pools, and in woods.

Fig. 184. — Pickering's Hylodes.

The American Toad is very useful to the farmer and gardener, in destroying insects.

SALAMANDERS, TRITONS, SIRENS, ETC.

Salamanders are Batrachians which have a long body and long tail, and which live upon the land, except when they go to the water to lay their eggs. There

Fig. 185. — Salamander.

are many kinds in North America, varying from three to twelve inches long. They are found mostly under

Fig. 186. — Triton.

stones, fallen trees, and rubbish. Tritons have nearly the same form, but live in the water. Tritons have the most wonderful power to repair or renew injured or lost parts. The legs may be cut off, and in less than a year they will grow again; and the limbs thus formed may also be cut off, and others will grow in their places; and even if the eye be destroyed another will grow to supply the loss.

In the Southern States is found the Congo Snake, an animal which is related to the Salamanders and Tritons. It is about two feet long, and lives in muddy waters. The Sirens have the gills in tufts, as in Figures 188, 189; thus even in the adult state they are like the young of Frogs and Toads. They live in the water.

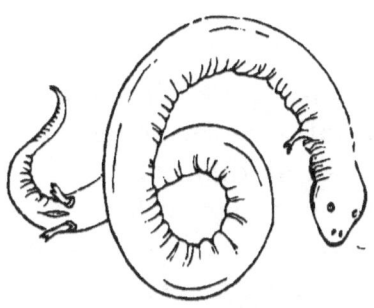

Fig. 187. — Congo Snake.

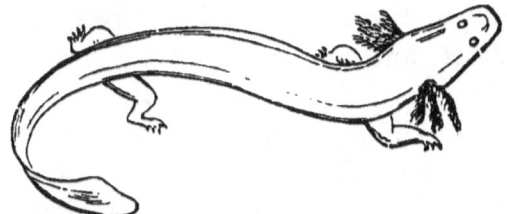

Fig. 188. — Mud Puppy.

Such are the Necturus, or Mud Puppy, of our Northern

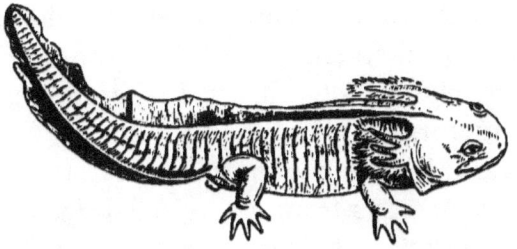

Fig. 189. — Axolotl.

Lakes, and the Siren, of the Southern States. The Axolotl of Mexico and our Western States is the young of a kind of Salamander. It usually does not complete its transformations, but remains permanently in the water, breathing by gills.

FISHES.

Fishes are Vertebrates which have cold blood, live wholly in the water, and breathe by means of gills. Most of them are scaly, but some are covered with a smooth skin, others have spines, and others still are covered with bony plates. The jaws are generally armed with teeth, and, in many cases, all parts of the mouth also, and even the gullet. Their movements are usually rapid, and their forward motion is mainly produced by the movements of the tail. The parts which correspond to the arms and legs of Quadrupeds are very short, and are called fins; and their use is mainly to balance and direct. The flesh is light-colored or white. In general, the eye of Fishes has little motion, and the pupil is always of the same size, both in light and darkness; the ear is wholly inclosed by the bones of the head. They are very voracious, feeding mainly upon smaller fishes, and other small animals, which they usually swallow whole. Those which feed on shellfish crush their food by means of the teeth in the gullet. Most Fishes lay eggs; a few kinds bring forth living young. Nearly all seem to have no care for their young, but eat them greedily. The number of eggs from a single fish in one season is often very great; the Salmon sometimes lays twenty thousand, the Cod more than nine million. The colors of Fishes are very beautiful, exhibiting metallic lusters, the brilliancy of precious stones, and the

SPINE-FINNED FISHES.

Fig. 190. — Yellow Perch.

Fig. 191. — Bream.

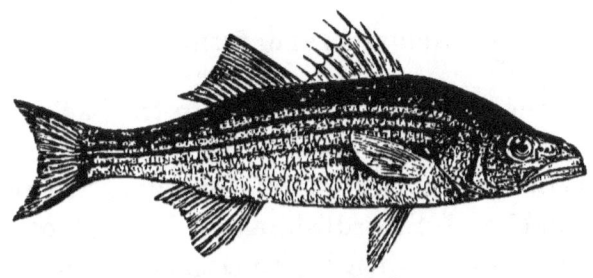

Fig. 192. — Striped Bass.

Fig. 193. — Stargazer.

Fig. 194. — Sea Robin

Fig. 195. — Stickle-
back.

Fig. 196. — Darter.

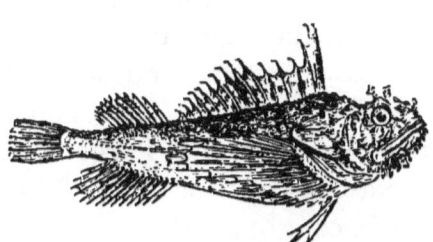

Fig. 197. — Sea Raven.

delicate tints of flowers; they are indeed the gems of the waters, as the Humming Birds are the gems of the air. The wonderful power and swift motion of some, the wholesome and delicious food furnished by many, and the exciting sport of their capture combine to render Fishes objects of great interest. The number of known kinds is about ten thousand.

Spine-finned Fishes.

Spine-finned Fishes have spines in the back or dorsal fin, and often in the lower fins. The Perch, Sea Bass, Pondfish or Bream, Stargazers, Sculpins, Sticklebacks, Porgee, Mackerel, Swordfish, and a host of others belong to this group, for it is the largest of all.

The American Yellow Perch, of our ponds and rivers, is known to every boy. The Striped Bass is caught in the sea near the shore, and the largest weigh seventy-five pounds each. The Pondfish or Bream is found in every pond, and the round cavities which it makes for its nest may be seen in great numbers near the shore. The Stargazers live in the sea, and have the eyes on top of the head, so that they appear as though looking at the heavens. The Sculpins live in the sea, and are often called Sea Robins, Sea Ravens, etc. The Sticklebacks are very small Fishes which inhabit both the sea and streams, and are very active and greedy, a single one having devoured seventy-five young fish in less than half a day. They construct very curious nests. The Weakfish and Porgee live in the Atlantic Ocean, and are caught for food. The Mackerel lives in the sea, and is caught on the coast of New England in immense numbers. The Swordfish has the upper jaw very

SPINE-FINNED FISHES. 127

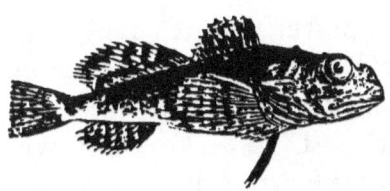
Fig. 198.—Sculpin.

Fig 199.—Scupaug.

Fig. 200.—Weakfish.

Fig. 201.—Mackerel.

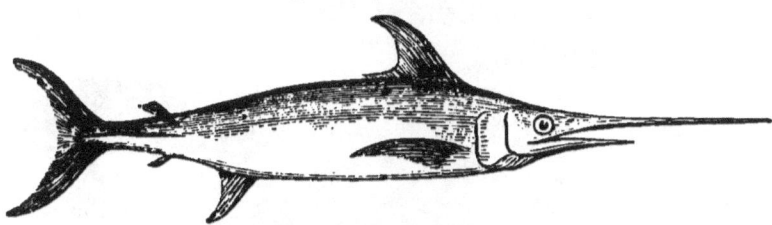
Fig. 202.—Swordfish.

Fig. 203.—Pilot Fish.

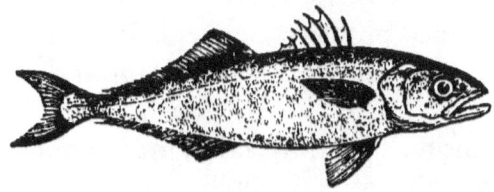
Fig. 204.—Bluefish.

much extended, forming a powerful and dangerous weapon, with which it attacks Whales and other large animals of the sea. The Bluefish is found in nearly all seas, and makes excellent food. The Dolphin lives in the Mediterranean, and in the Atlantic, and is celebrated for its beautiful colors, and for the brilliant tints which it has when dying. The Surgeon has a sharp spine or lancet on the side of its tail; it lives in the sea. Mullets are small fishes which live in the sea, and in fresh waters. Eelpouts are long, somewhat eel-shaped fishes, which the fishermen catch when fishing for Cod. The Goosefish of the Atlantic, is large, sometimes weighing seventy pounds, and has such a big mouth that it swallows fishes almost as large as itself. Gulls and other sea birds are often found whole in its stomach. The Toadfish, of the Atlantic, is about a

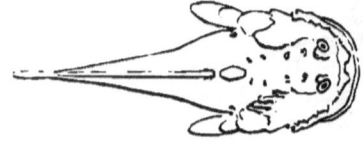

Fig. 205. — Toadfish.

Fig. 206. — Conner.

foot long, and seems to care for its young. The Conner is very abundant on the coast of New England.

Soft-finned Fishes.

These Fishes have no spines in their fins. They are the Carp, Dace, Shiners, Suckers, Pike, Pickerel, Garfishes of the sea, Flying Fishes, Salmon, Herring, Cod, Eels, etc.

The Common Shiner, found in most ponds, lakes, and rivers, is from three to six inches long, and of a golden color. The Pickerel, so well known in the fresh waters,

SPINE-FINNED FISHES.

Fig. 207. — Blunt-nosed Shiner.

Fig. 209. — Surgeon.

Fig. 208. — Mullet.

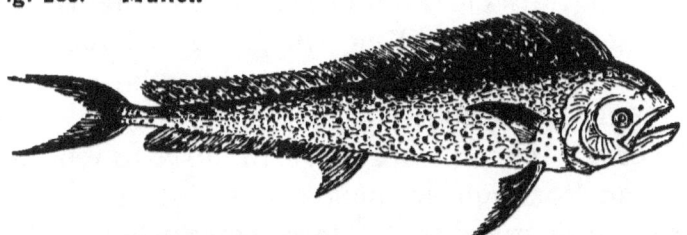

Fig. 210. — Dolphin.

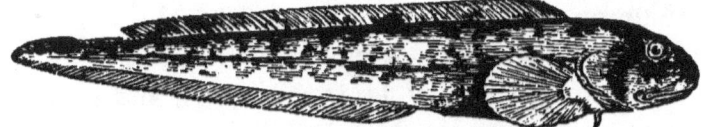

Fig. 211. — Eelpout.

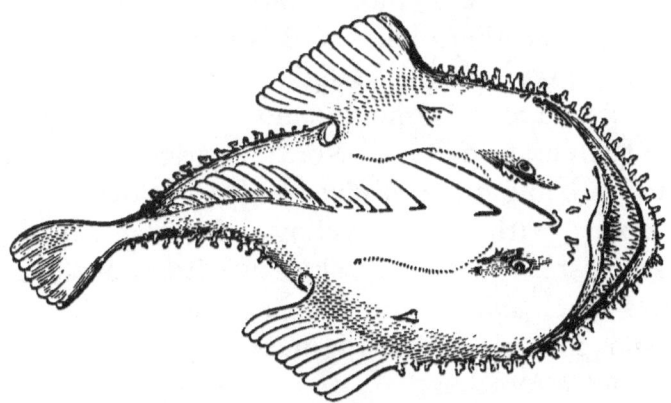

Fig. 212. — Angler, or Goosefish.

is a handsome fish, of fine flavor, and the sport of its capture is very exciting. The Garfish lives in the sea, and has an extremely long head and body; the jaws are pointed, and armed with many small teeth, and its bones are green. Flying Fishes have the fins, which are directly behind the gills, so large that they are able to sustain themselves in the air for a few moments, thus appearing to fly. They live in all warm and temperate seas, and are from three inches to a foot in length. The Blindfish is found in the waters of the Mammoth Cave, Kentucky, and is about three inches long. Its eyes are under the skin, so that it is perfectly blind, and thus adapted to the dark waters of the cave. The Horned Pout, from six to ten inches long, and common in ponds and sluggish streams, has the head armed with sharp spines, which inflict a smarting wound on the hand of the careless fisherman. The Salmon is a most beautiful fish, whose home is in the Arctic seas, but it comes southward and ascends rivers for the purpose of laying its eggs, and is caught in large numbers. Its flesh is delicious, and it weighs from ten to thirty pounds or more. The Lake Trout inhabits our northern lakes, and is from two to five feet long, of a gray color with lighter spots. The Brook, or Speckled, Trout, is found in most of the clear streams of the temperate parts of North America, and is very beautiful, being dark above, silvery below, and the sides dotted with red and yellow. Its flesh has a very delicate taste. It is very shy, and its capture often requires much skill. The Herring lives in the Arctic seas, and comes southward in spring to lay its eggs. It is about a foot long.

The Cod inhabits the north Atlantic, and attains a

SOFT-FINNED FISHES.

Fig. 213. — Pickerel.

Fig. 214. — Shiner.

Fig. 215. — Flying Fish.

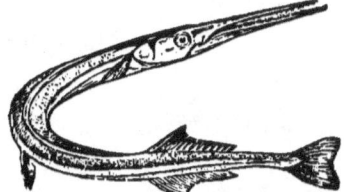
Fig. 216. — Garfish.

Fig. 217. — Blindfish.

Fig. 218. — Horned Pout.

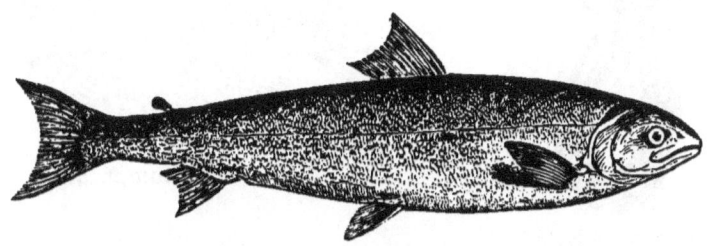
Fig. 219. — Salmon.

Fig. 220. — Herring.

Fig. 221. — Speckled Trout.

Fig. 222.—American Cod.

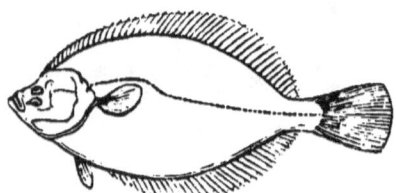

Fig. 223.—Flounder.

Fig. 224.—Burbot.

Fig. 225.—Eel.

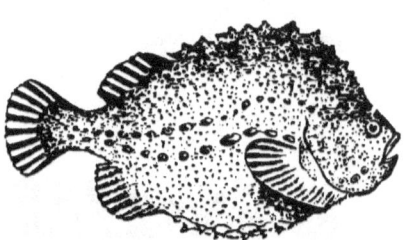

Fig. 226.—Lumpfish.

Fig. 227.—Remora.

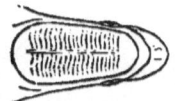

Fig. 227, a.—Top of head of Remora.

weight of even a hundred pounds in some cases. It is taken in immense numbers on the Banks of Newfoundland, and when salted and dried is carried to all parts of the world. The Flounders are marine Fishes which have the body flattened on the sides, and both eyes are on the same side of the head. The side upon which the eyes are placed is always uppermost, and is dark colored, while the opposite side is white. They swim, therefore, on one side, and they keep close to the bottom. Flounders are from six inches to two feet long, and are caught in great numbers, even from the wharves. Halibuts are shaped like the Flounders, and in some cases weigh six hundred pounds. The Flounders and the Halibuts are the only backboned animals which have the right and left sides unlike. The Lumpfishes are those whose ventral fins are so joined as to form a sort of cup, by which they are able to attach themselves firmly to rocks or other objects. Pennant, the naturalist, says that he put one into a pail of water, and it adhered so tightly to the bottom that he lifted the whole pailful by taking hold of the fish by the tail. It lives in the north Atlantic. The Remora has a flattened head, so constructed that the fish can attach itself by it to other marine animals, such as Sharks. It is a foot or more in length. Eels have a long, round body, covered with a thick, soft skin, live in both fresh and salt waters, and keep near the bottom, often lying concealed in the mud.

Tuft-gilled Fishes.

These Fishes have their gills in tufts, and are known as Pipefishes and Sea Horses, on account of their singular forms. Pipefishes have a very long and slender body covered with hard plates, and a long snout with the mouth at the end. They live in the warm seas. After the eggs are laid, the male takes them in a sort of sack and carries them about with him till they are hatched. Sea Horses have a short body covered with spiny plates, a tail adapted for grasping objects, and the head and neck resemble those of a Horse. They are from three to six inches long, and live in the sea.

Puffers, Trunkfishes, etc., or Plectognathi.

Puffers have the body covered with spines, and can swell themselves like a balloon by swallowing air. The Common Puffer lives in the Atlantic Ocean, and is about a foot long.

The Sunfish, of the Atlantic, grows to the length of four feet, and weighs five hundred pounds.

The Trunkfish has the head and body covered with bony plates, so firmly attached to each other that they form a shield, and the mouth, tail, and fins are the only movable parts. Two or three kinds are found on the Atlantic coast of the United States.

Sturgeons and Garpikes.

Sturgeons are Fishes whose skeleton is a sort of cartilage, instead of being bony, as in those already described. They are also covered with bony plates placed in rows along the whole length of the body, and the mouth is under the snout, and can be much protruded. They are from three to ten feet long, inhabit lakes and the ocean, and ascend rivers. See Figure 233.

TUFT-GILLED FISHES, PUFFERS, ETC.

Fig. 228. — Sea Horse.

Fig. 229. — Pipefish.

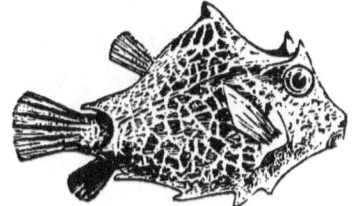

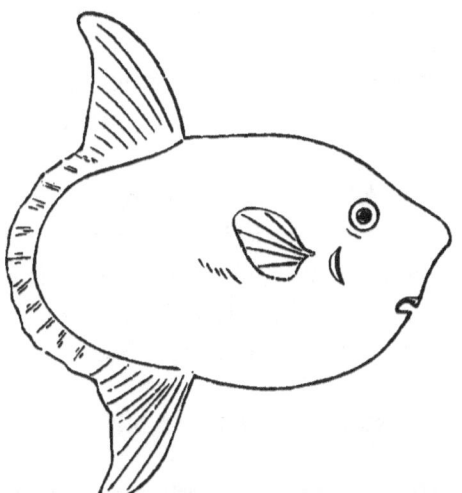

Fig. 230. — Puffer.

Fig. 231. — Trunkfish.

Fig. 232. — Sunfish.

Fig. 233. — Sturgeon.

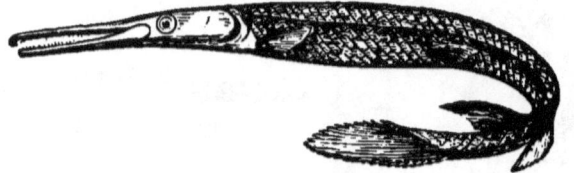

Fig. 234. — Garpike.

The Garpike of the Mississippi valley has a long body and long jaws, armed with numerous sharp teeth. Its body is covered with enameled scales fitted closely together.

Sharks, or Selachians.

These are marine Fishes with a cartilaginous skeleton. They are frequently large, and usually very ferocious. They vary from four to thirty feet in length; their teeth are numerous, sharp as lancets, and inflict the severest wounds. The smaller marine animals, and even men, fall a prey to them.

The Rays, or Skates, are broad and flat, from two to six feet or more in length and width. Those called Vampires are sometimes sixteen feet wide, and weigh several tons. One kind, the Torpedo, gives violent electrical shocks when touched. See Figures 241, 242.

Suckers, or Cyclostomi.

The true Suckers are the least perfect and lowest of all the Fishes, and their tongue moves forwards and backwards like the piston in a pump, enabling them to produce a vacuum, and thus to fix themselves to other fishes. The Sea Lamprey, two or three feet long, the Hagfish or Myxine, six or eight inches long, and the Brook Lamprey about a foot long, are of this kind. The

SHARKS.

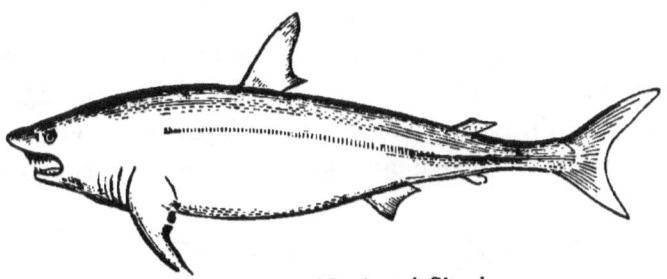

Fig. 235.— Mackerel Shark.

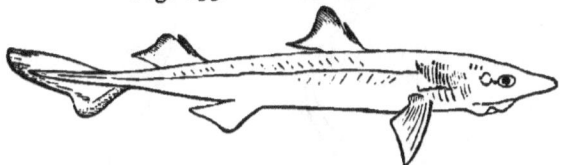

Fig. 236.— Dogfish Shark.

Fig. 238.— Head of Hammerhead Shark.

Fig. 237.— Head of Mackerel Shark.

Fig. 239.— Hammerhead Shark.

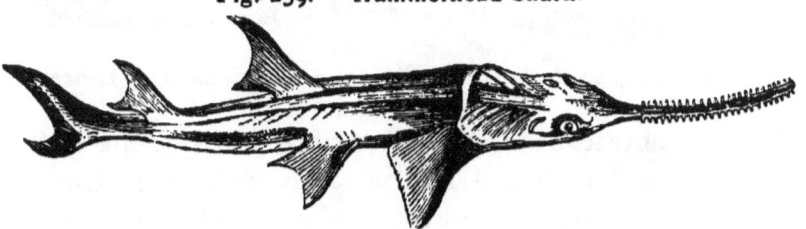

Fig. 240.— Sawfish — a Shark.

138 VERTEBRATES: FISHES.

Lamprey ascends rivers, and piles up heaps of stones, among which it lays its eggs. See Figures 243, 244.

The Amphioxus, or Lancelet, is about two inches long. It is the lowest and simplest in structure of all

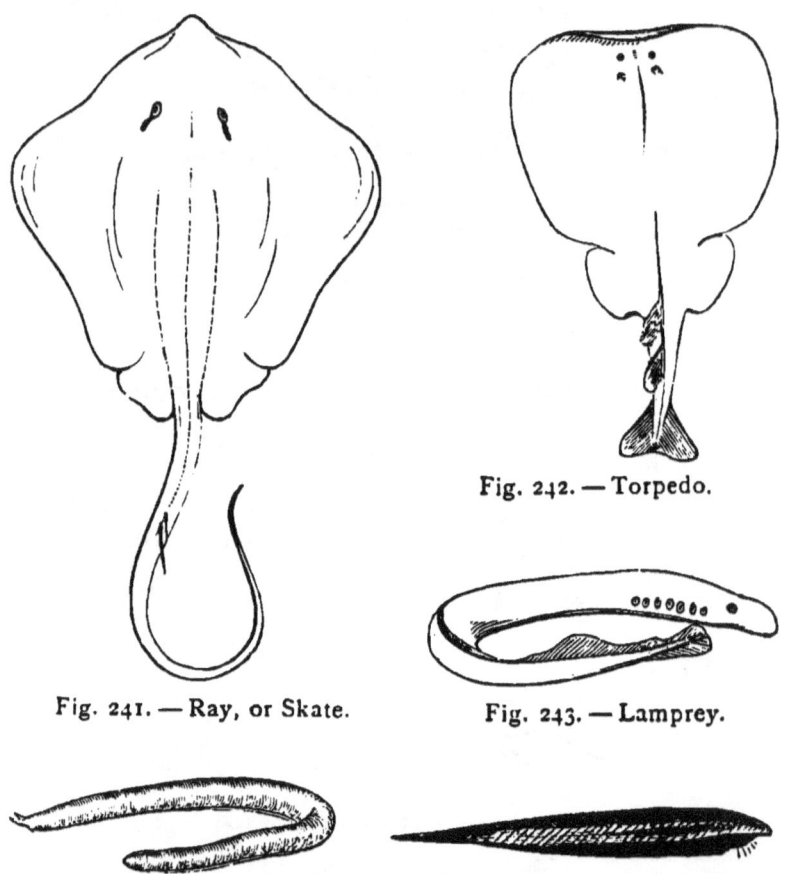

Fig. 241. — Ray, or Skate.

Fig. 242. — Torpedo.

Fig. 243. — Lamprey.

Fig. 244. — Hagfish, or Myxine. Fig. 245. — Amphioxus, or Lancelet.

the Vertebrates. It is a beautiful, transparent little creature, and is found in the sandy bottom of Chesapeake Bay and in similar shallow, quiet waters in tropical and sub-tropical seas. See Figure 245.

TUNICATES.

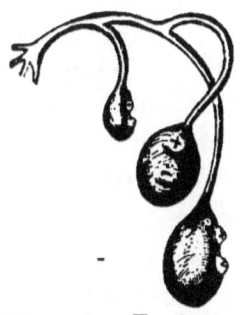

Fig. 246.—Tunicate.

These animals have no shell, but are covered with a tough tunic, or skin. Sometimes they grow in clusters, attached by a stem to seaweed, rocks, or floating timber. They vary from the size of a pea to several inches in diameter. They are sometimes called Ascidians, from a word which means a *leather bag*. Strange as it may appear they are not distantly related to the Vertebrates, as is shown by their development.

ARTHROPODS, OR JOINTED ANIMALS.

Arthropods have no internal skeleton; the hard parts are exterior; the body is made up of a series of similar rings, each of which bears a pair of jointed legs. They include the Insects, — Bees, Butterflies, Flies, Beetles, Bugs, Grasshoppers, Darning Needles, etc.,— Spiders, Scorpions, Mites, Lobsters, and Shrimps.

INSECTS.

Insects breathe by means of air holes along the sides of the body, the openings of branching air tubes which carry air to every part. The term Insect means *cut into;* the animals seem to be cut into, or jointed. The body is divided into three parts,—the head, middle body or thorax, and hind body or abdomen. On the head and near the eyes are two jointed members, called antennæ, supposed to be connected with the sense of smell or of touch, or of both; to the thorax are attached the legs and wings; and the hind body contains the organs of digestion, and often has a sting, or piercer. Insects either bite their food or suck it. Those which bite their food have an under and upper lip, between which are two pairs of jaws which move sidewise, and two pairs of little feelers, which they use to touch and examine the food. Those Insects which suck their food have either a long tube, as Butterflies and Moths; a piercing sucker, as Mosquitoes; a softer one, used for lapping, as Flies; or a jointed one, which is doubled under the breast when not in use, as Bees. The eyes of Insects appear to be only two in number, but each is composed of many single eyes,— often thousands, and in some cases the astonishing number of twenty-five thousand,— closely united. Many Insects have also one, two, or three single eyes on the crown of the head. The legs are six in number, and are attached to the under side of the thorax; the wings are four, or sometimes two, and vary greatly in form and thickness, in veinings, and in the manner of folding when at rest. The hind body is the largest portion, and most of the air holes are found in it.

Insects are produced from eggs. A very few do not lay their eggs, but retain them in the body till hatched; others always lay their eggs where the young will find a plentiful supply of food. Many Insects undergo great and wonderful changes in form and habits; so great, that the same insect, at different ages, might be taken for as many different animals. For example: a caterpillar, after feeding upon leaves until it is fully grown, casts off its skin, and appears as a much smaller, oval body, which neither moves about

Fig 247. — Larva.

Fig. 248. — Pupa of Fig. 247.   Fig. 249. — Imago of Figs. 247, 248.

Fig. 250. — Larva.

Fig. 251. — Pupa of Fig. 250.

Fig. 252. — Imago of Figs. 250, 251.

nor takes food. After remaining awhile in this state, the skin bursts open, and there comes forth a Butterfly or a Moth, whose wings expand, and harden, and are soon able to bear it away in search of flowers, upon whose honey it feeds. In its first state it is called a *larva*, — a word which means a *mask*, — because its future form is masked or concealed; in the second state it is called a *pupa*, — a word meaning *infant*, — from a slight resemblance that some insects in this state bear to an infant clothed with bandages, according to a custom among the Romans; and it is also often called a *chrysalis*, from a Greek word which means *gold*, because some of the pupæ are adorned with golden spots; in the third state it is called a perfect insect, or *imago*, from a word which means *image*, because the image concealed in the skin of the pupa has come forth. These different states are plainly shown on page 141. Some caterpillars spin a silken covering, which is called a *cocoon*, from a word which means a *shell;* all the silk of the world comes from the cocoons of these little creatures. Insects which pass through the changes just described are said to undergo a complete transformation; but there are some insects which do not change their form so completely. Grasshoppers, for instance, are active during their whole lives, never passing through an inactive pupa state. When hatched from the egg they have legs, but no wings; later their wings begin to grow, and, at length, having shed their skin several times, each time appearing with longer legs and more perfect wings, they reach their full growth, shed

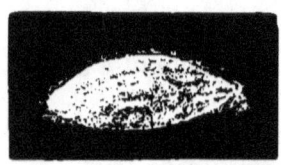

Fig. 253. — Cocoon.

the skin for the last time, and appear as perfect, or adult, grasshoppers. Such Insects undergo only a partial transformation.

Insects are the most numerous of all the classes of animals, there being more than two hundred thousand kinds described; while the undescribed forms are far more numerous. And the study of Insects is one of the most interesting and fascinating in which one can engage. The study of Insects is also very important, that we may know which are injurious to the farm, orchard, garden, granary, and closets, and by knowing their habits be able to resist their attacks; and that we may know which are of use to man: for the Bee gives us delicious honey; some of the Beetles are of use to the sick; some of the little Bark Lice, as the Cochineal, yield rich dyes; and some of the Caterpillars furnish all the world with silk.

Bees, Wasps, Ichneumons, etc., or Hymenoptera.

These Insects have four wings which are more or less transparent, the hind pair being the smaller, and all with a few branching veins. They have two pairs of jaws,— the upper pair fitted for biting, while in the Bees the lower pair with the lower lip is adapted for collecting honey. The females have either a sting or a piercer for laying their eggs. They surpass all other Insects in the number and variety of their instincts. The word Hymenopter means *membrane-winged*.

Bees.

Bees have a hairy body, and their lower lip is lengthened into a sort of proboscis, which is jointed and can be folded under the head; the first joint of the hind legs is often very large, and fitted for collecting and carrying the pollen of flowers.

The Hive or Honey Bee is originally from Asia, but has now spread over Europe and America. It is seen almost everywhere in hives, and it is also quite common in a wild state, and often far from human dwellings. In a wild state, Bees of this kind have their home in hollow trees and in clefts of rocks. In every nest or hive there are three kinds, a female or queen, males or drones, and workers. In a well-stocked hive there are two thousand males, fifty thousand workers, but only *one* queen. The workers are the smallest;

Fig. 254.—Queen.

Fig. 255.—Worker.

Fig. 256.—Drone.

Hive Bee.

they fly over the surrounding country and collect all the materials to form the structure called the comb; they build the cells and store them with honey; they feed and protect the young; they wait upon the queen; they do all the work of the hive. The males or drones have a thicker body, and no sting; they perform no labor, but are supported by the workers. The queen is much larger than the others, has a sting, and is the sole mistress of the hive. She lays all the eggs, and seldom goes out except to lead a swarm. The honeycomb is one of the most interesting of insect structures, and is arranged in the hive in the most regular manner. The cells are six-sided, and are built in just the shape to save all the room, to be the strongest, to contain the greatest amount of honey, and to require the least amount of wax in their construction.

There are certain cells in which the queen lays her eggs, depositing one in each cell; and when the eggs are laid, the workers fill the cells with the pollen of flowers mixed with water and honey, — this is food for the larvæ. In about two days the eggs hatch into small white larvæ, and in five or six days these begin to spin a cocoon, and soon go into the pupa state. A queen comes forth from this state in sixteen days, workers in twenty days, and drones in twenty-four days. As only one queen can live in a hive, whenever a young queen is hatched she is carefully guarded from the old one by the workers, till it is settled whether the old queen will be wanted to lead forth a swarm. If a new swarm is not to go forth, the old queen is allowed to approach the young queen and royal cells, and destroy the brood, with her sting. If the old queen leaves with a swarm, a young queen is set free and immediately endeavors to destroy the others, but is prevented by a guard of workers, while there is a prospect of another swarming; if she departs with a swarm, another queen is set free, and so on till further swarming is impossible; then the young queen is allowed to kill all her sisters. If two queens hatch at the same time, they instantly engage in conflict, the other bees favoring the battle, and when one is killed, the survivor is recognized as queen. When a hive loses its queen, there is the greatest confusion; after several hours they become quiet, and if there are no eggs or larvæ in the cells from which a new queen may be hatched, they become discouraged, cease to labor, and the whole colony soon dies. If there are eggs or larvæ in the cells, the bees select one,— the larva of a worker,— and destroying the cells adjoining, so as to make a royal cell, they supply the grub with the sort of food prepared for queens, and in this way soon raise another queen.

The Humble Bees are larger than the Hive Bees, and have very hairy bodies. There are more than forty kinds in North America. They build nests in the ground, or under stones, or in deserted mouse nests, and their cells are larger and egg-shaped. Sometimes there are four hundred bees in a community, the descendants of one female bee which survived the winter and founded the colony in the spring. The Carpenter Bees are large. They cut tubular holes in posts and stumps, and lay their eggs there in layers of pollen. The Mason Bees make their nests of sand, in crevices.

Wasps.

Wasps usually live in colonies composed of males, females, and workers. Unlike Bees, they prey upon

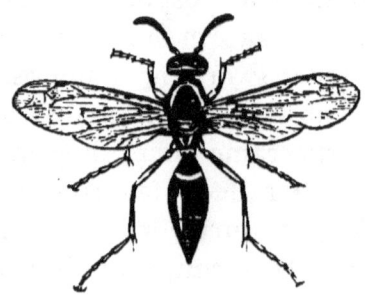

Fig. 257.— Wasp.

other insects. They build nests under ground, or in holes, or attach them to bushes, trees, fences, or buildings. The nest is usually made of a substance which they gnaw from wood, and which, by the action of their jaws, they reduce to a pulp, which hardens into a sort of paper. The Wasps were the first paper-makers, and they were the first to show that paper can be made of wood. The combs lie horizontally in the nest, are made of the same paper-like material as the nest, and each is attached to the one below it by a sort of pillar. The cells contain no honey, but are built for places in which to rear the young. The colony is dissolved on the approach of winter, the males die, and the females seek a sheltered winter retreat. Each female that survives the cold founds a new colony in the spring, building a few cells and laying her eggs, from which are hatched only workers. These assist the parent, and at length, in autumn, three generations have been produced, the last composed of males and females, and the nest has grown from a few cells to one containing thousands. The Hornet is one of the largest of the Wasps, and was brought to this country from Europe.

Some kinds of Wasps build open nests of a few cells, and attach them to some object by a short stem. Other kinds build their nests of mud, and store them with insects for the food of the larvæ; these are the Mud Wasps. They have the hind body joined to the thorax by a long stem or pedicel, and their color is shining blue, or black, or black and orange, or brown and red. One of the black and orange Mud Wasps built two beautiful mud cells in the corner of my room. She worked very industriously and rapidly, building a cell in a few hours. Flying in at the open window, with a ball of mud in her mouth, she moved quickly around the room, then flew up to the spot where she was building, and depositing her mud, shaped it with her jaws with all the care and regularity of a perfect mason. The day after she finished the first cell, she filled it with spiders and sealed it over with mud. On opening it to examine the insects stored within, quite a large hole was accidentally made; this she very soon discovered, and began to repair it, and in about five minutes she had completely closed it. The second cell was soon sealed like the first. Fig. 258 shows them, as they appeared before the second was filled with spiders and closed.

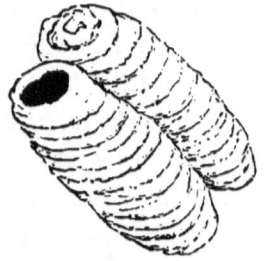

Fig. 258. — Mud Wasp's nest.

Ants.

Ants live together in colonies, which are often very large, and made up of males, females, and workers. The workers have no wings, but the males and females have wings, and the females have the power of throwing them off. Some kinds of Ants make their nests in the ground; others raise large ant-hills; and others live in stumps and trunks of trees. The workers take care of the nest and rear the young; they go abroad in search of food, communicate with and assist each other, feed the larvæ, take them into the sunshine in fair weather and back again on the approach of a storm or at night, and watch over them earnestly and faithfully. Ants are fond of sweet things, and some obtain such food from the secretion of aphides, or plant lice, — little insects which live upon the juices of plants, and yield a honey-like fluid. Some kinds of Ants collect large numbers of aphides and keep them on plants, that they may eat the sweets which they produce. There is generally but one species of Ant in each nest, but in some cases the workers, or, rather, warriors, make slaves by visiting the hills of other species, forcibly taking the larvæ and pupæ, and bringing them back, where they are tended and reared by workers of the same kind which have before been stolen in the same way. Ants are very warlike, and engage in deadly pitched battles.

Ichneumons.

These Insects have a long, hard, slender body, long antennæ, and the ovipositor is usually long; the latter

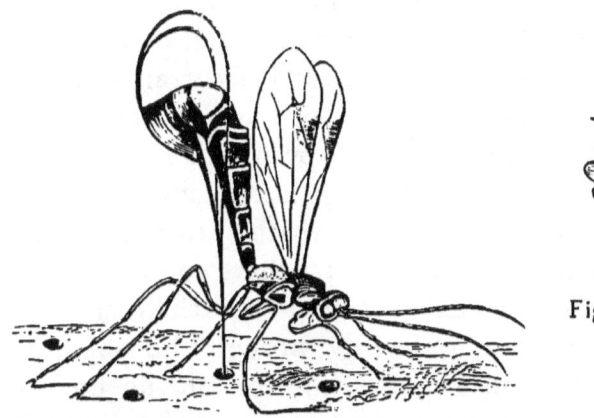

Fig. 260. — Ichneumon.

Fig. 259. — Ichneumon laying her eggs in holes bored by the Boring Sawfly, Figure 262.

is sometimes two or three times the length of the body. They lay their eggs in the eggs, larvæ, and pupæ of other insects, and thus destroy great numbers of them. Sometimes the eggs are laid upon the outside, but usually inside. When laid on the outside of the pupæ, the Ichneumon, as soon as hatched, eats its way into its victim; when laid inside, it feeds upon the body but attacks no vital part, and the insect does not die till the Ichneumon is ready for the pupa state.

Gallflies.

These are very small Insects, and the females have a long, slender ovipositor, with which they insert their eggs into leaves and other parts of plants. These punctures cause outgrowths called galls, which vary in size, form, and solidity, according to the nature or part of the plant that is wounded, and according to the kind of Gallfly that makes the wound. Some are shaped like an apple, as the gall of the oak; some like a bunch of currants; some are almost as hard as iron; and some are juicy and pulpy, like fruit. At length the eggs hatch, and the larvæ feed upon the vegetable matter which surrounds them. Some galls have only one tenant, others contain many, and usually these Insects undergo all their changes within the galls, and, gnawing through the shell, fly away; but some kinds gnaw through at the end of their larva life, and enter the ground to go into the pupa state. The nutgalls used in making ink, in coloring, and in medicine, are caused by an insect which punctures a species of oak common in western Asia. The Rosebush Gallfly punctures the stems of rose-

Fig. 261. — Rosebush Gallfly.

bushes, producing woody galls. One of the largest species is the Willow Gallfly; its galls are found at the end of basket willow twigs.

Boring Sawflies.

The Boring Sawflies have a long body; the hind body is blunt and ends in a horny point. Extending from beneath the hind body is a long, saw-like, and

Fig. 262. — Boring Sawfly, or Pigeon Tremex.

powerful borer, used to make holes in trees, in which to lay their eggs. The larvæ live in tree-trunks.

True Sawflies.

The females of the true Sawflies have an ovipositor composed of two saws, enclosed between two horny pieces, which serve as a sheath. These saws are projected and withdrawn when the insect is cutting a place

Fig. 263. — Fir-tree Sawfly. Enlarged.

for her eggs; but not together, for while one is pushed forward, the other is withdrawn. When the hole is cut deep enough, the egg is deposited within. Saw-flies are sluggish, and fly only on the warmest days. The larvæ are found together in large numbers on the leaves of the birch and alder. When disturbed, they take very curious attitudes, appearing to stand upon the head, curling into an S, or coiling with the head in the center looking somewhat like a snail-shell.

BUTTERFLIES AND MOTHS, OR LEPIDOPTERA.

The word Lepidopter means *scaly-winged*, and is given to these Insects because their wings are covered on both sides with minute scales. These are removed by the slightest touch, and to the naked eye look like a mealy powder; but when seen under a microscope, they are found to be little scales attached to the skin by a short stem. The tongue is long, and adapted for suction; when not in use it is rolled up like a watch spring beneath the head, and partly concealed on each side by a little feeler. They have six legs, the first pair being short, and, in some cases, folded under the breast; the feet end in a pair of claws. The young of Butterflies and Moths are called caterpillars, and these have from ten to sixteen legs. Six of the legs correspond to those of the Butterfly. The rest are unjointed projections of the abdomen, and are called prolegs or proplegs. Most caterpillars feed upon the leaves of plants. Some eat buds, blossoms, seeds, and roots, and others eat the solid wood. Some eat woolens, others furs, others meat, lard, wax, and flour. Some kinds herd together in great numbers, and build nests in which they live, or to which they retire for shelter; others live in solitude, either in the light and air, or sheltered in leaves folded over them, or in silken sheaths which they make; and some conceal themselves in the ground, coming forth only to eat. In the middle of the lower lip there is a little tube, from

Fig. 264.—Turnus Butterfly.

which the caterpillar spins silken threads. Two long slender bags within the body, ending in the spinning tube, contain the material from which the silk is made;

these correspond to the salivary glands. The silk is a sticky fluid, which hardens into a thread as soon as it comes to the air. Some caterpillars spin but little silk, others produce it in abundance.

Caterpillars change their skins about four times in coming to their full growth as caterpillars; and when about to change into the pupa or chrysalis state, they cease eating, and many of them spin around their body a silken covering called a cocoon, others suspend themselves by silken threads without making a cocoon, and others enter the ground. When the caterpillar is prepared for the change, it bursts the skin on the back, draws out the forward part of its body, and works the skin backward until it throws it off; and now it is a chrysalis, shorter than the caterpillar, and at first sight it appears destitute of head and limbs; but on looking more carefully we perceive traces of head, tongue, antennæ, wings, and legs. Some chrysalides are angular, but most of them are smooth, rounded at one end, and tapering at the other; they remain motionless, or only move the hind part of the body when touched. At length, the inclosed insect is ready to come forth, and by many movements its bursts the skin of the back, and the Butterfly or Moth appears. At first it is soft, weak, and moist, with small and shriveled wings; but soon the moisture passes off, the limbs become firm, the wings expand, and the perfect and beautiful insect flies away to feed upon water and the honey of flowers. Butterflies and Moths do not increase in size; they are full grown when they emerge from the pupa skin; and after having laid their eggs, they soon die. Butterflies fly in the daytime, have their wings erect when at rest, their antennæ are threadlike, with a little knob at the end, and their larvæ do not spin cocoons. Moths fly mainly at night, have their wings when at rest more or less sloping like a roof, and their antennæ are variously formed, but never knobbed at the end.

Papilio Butterflies.

The Turnus Butterfly is one of the largest in North America. Its color is a beautiful yellow, with black markings, and the hind wings are tailed. The caterpillar feeds upon the leaves of the apple and wild

Fig. 265. — Larva of Asterias Butterfly.

Fig. 266. — Pupa of Fig. 265.

Fig. 267. — Asterias Butterfly.

cherry trees, folding them up so as to make a case for itself. When fully grown, it is about two inches long, green above, with rows of blue dots and yellow and black marks, and its head and legs are pink. It becomes a chrysalis early in August, and comes out a butterfly the next summer. See Figure 264.

The Asterias Butterfly is black, with two rows of yellow dots on the back, and two rows of yellow spots across the wings; the hind wings are tailed, and have seven blue spots between the two rows of yellow ones, and an eye-like spot of an orange color with a black center. The female is much larger, and has fewer yellow spots on the upper surface. The caterpillar lives upon such plants as the carrot, parsnip, celery, and anise. It is green, with a band made up of yellow and black spots on each ring. When touched, it thrusts out from the head a pair of soft, orange-colored horns. These have an unpleasant odor, which makes the caterpillar disagreeable to birds. Thus it escapes being eaten. In July it reaches its full growth as a caterpillar; then it seeks a sheltered spot on the side of a building or fence, spins a tuft of silk, fixes its hind feet in it, then makes a loop of silk, and, passing its body through it, rests upon it as a support; soon it casts its caterpillar skin and becomes a pupa or chrysalis, Figure 266. In about a fortnight the butterfly, Figure 267, appears.

White and Yellow Butterflies.

The Philodice, or Yellow Butterfly, expands about two inches, and is common in fields and by roadsides throughout the summer. The White Butterfly, or Pieris, is of about the same size, and is also common.

Nymphalis Butterflies.

These Butterflies are remarkable for their beautiful colors. The Misippus Butterfly has the wings tawny yellow, veined with black, and a black border spotted with white, the fore wings have near their tips a black patch spotted with white, and on the hind wings is

Fig. 268. — Misippus Butterfly.

a curved black band. The caterpillar is pale brown, marked with white on the sides, and on the second ring are two blackish horns. The butterfly is seen in June and September.

Satyrus Butterflies.

These have their wings broad and rounded, and those called Hipparchians have the wings of a delicate brown color, with beautiful eye-spots. They are very common in groves and about bushes late in the summer. Closely related to these is the Mountain Butterfly, which is found only on the top of Mount Washington.

Fig. 269. — Mountain Butterfly.

Skippers.

Skippers are butterflies which have a short body, large head, and large eyes; and the antennæ have the knob at the end either curved like a hook or ending in a little point bent to one side. They are called Skippers because they fly with a jerking motion. They are generally of a rich brown color, marked with spots of yellow, and expand from an inch and a half to two inches and a half.

Fig. 270. — Skipper.

The Tityrus Skipper is one of the largest and most beautiful species. Its wings are brown; the forward wings have a yellow band across the middle and yellow spots near the tips, and the hind wings have a broad, silver-colored band across the middle of the under side. It is found about clover and other flowers in June and July. The females lay their eggs on the leaves of the locust trees. The caterpillar, when full grown, is about two inches long, pale green, with cross streaks of darker green; the head and neck are red, with a yellow spot on each side of the mouth.

Hawk Moths, or Sphingidæ.

These Moths are large, and have the antennæ thickest in the middle and usually hooked at the tip, and the wings long and narrow. During the morning and evening twilight, they may be seen flying from flower to flower with great swiftness, and are easily mistaken for Humming Birds. A few kinds fly by day and in bright sunshine. The caterpillars are very large, and are remarkable for their curious attitudes, which reminded Linnæus of the Sphinx, a sculptured monster of the Egyptians.

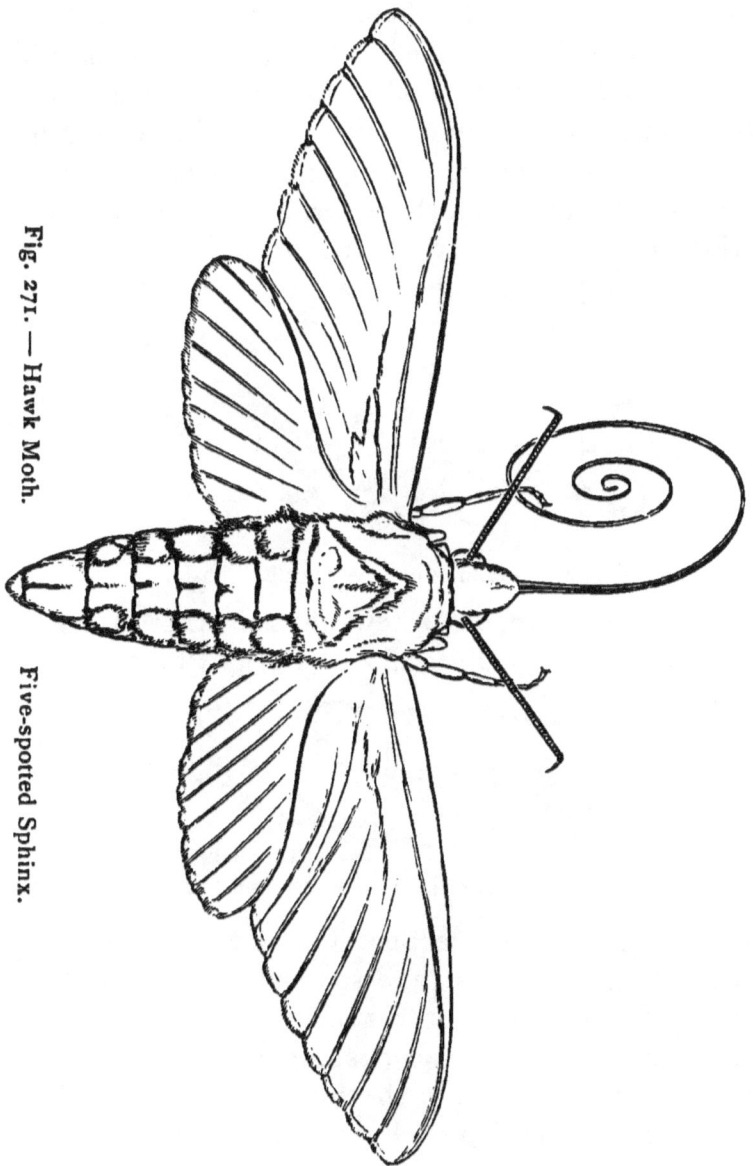

Fig. 271. — Hawk Moth. Five-spotted Sphinx.

The Five-spotted Sphinx expands about five inches, and is of a mixed grayish and blackish color, and on each side of the body there are five orange-colored spots surrounded by black. Its tongue, when fully unrolled, is five or six inches long, but when not in use is coiled up nearly out of sight. The caterpillar is known as the potato worm, and is green, with oblique whitish stripes on the sides, and a thorn-like projection on the tail. It attains its full length, three inches or more, in August, and then buries itself in

Fig. 272. — Larva of Five-spotted Sphinx.

the ground. Here, in a few days, it throws off its skin and becomes a chrysalis, of a bright brown color,

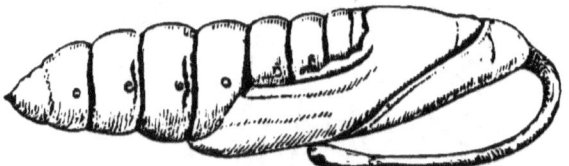

Fig. 273. — Chrysalis of Five-spotted Sphinx.

with a long tongue-case bent over from the head, its end touching the breast, and somewhat resembling the handle of a pitcher. It remains in the ground all winter, and in the following summer the large moth crawls out of it, comes to the surface, mounts a plant, and waits till the approach of evening, when it flies away in search of food.

Clear-winged Sphingidæ, or Sesias.

These are known by their transparent wings and

Fig. 274. — Clear-winged Sesia.

broad tails. They are seen in the daytime hovering over flowers, and are very beautiful.

Peach-tree Borers.

The Peach-tree Borer, in its winged form, resembles a Wasp. The general color is steel-blue, with yellow markings, and the male has all the wings transparent; but the female has the fore wings blue and opaque. The eggs are laid upon the trunk of the tree, near the roots. When hatched, the larvæ bore into and devour the inner bark and sap wood. When about a year old they become chrysalides, and come forth as moths from June to October.

Fig. 275. — Peach-tree Borer.

Silkworm Moths.

These Moths have the head small, antennæ generally feathered or toothed, tongue short, thorax woolly, and the fore legs hairy. Most of the caterpillars spin cocoons. Some of these Moths are small, and others are the largest of the Lepidoptera.

One of the most elegant kinds is the Beautiful Deïopeia. Its fore wings are yellow, crossed by white

Fig. 276. — Beautiful Deïopeia.

bands, on each of which is a row of black dots, and the hind wings are scarlet with an irregular black border.

The Salt-marsh Moth expands about two inches; the fore wings are white, hind wings and hind body yellow;

Fig. 277. — Larva of Salt-marsh Moth.

Fig. 278.— Pupa of Salt-marsh Moth.　　Fig. 279.— Salt-marsh Moth.

the wings are spotted with black, and the hind body has a row of black spots above, a row below, and two rows on each side. The female has all the wings white, or all light gray, with the black spots.

The Common Silkworm is celebrated as the Insect which produces the greater part of all the silk used in the world. It is the larva, or caterpillar, of a Moth, — *Bombyx mori*, — which expands about two inches, and which is of a light color, with two or three obscure streaks, and a spot on the upper wings. It feeds upon the leaves of the mulberry tree, and spins a cocoon about an inch and a half long, of a yellow color, and which contains about one thousand feet of silk. This Silkworm is a native of China, but is now raised extensively in Europe, and, to some extent, in this country. The larvæ of several other moths, most of them of large size, are now raised, not only in Asia, but also in Europe and in the United States, for the purpose of producing silk.

The Cecropia Moth, the Promethea Moth, the Luna Moth, and the Polyphemus Moth are all large and magnificent species, — the largest in North America. They have the antennæ broadly feathered on both sides, and beautiful eye-like spots on the wings. All but the Promethea expand five or six inches, and the latter expands about four inches. They appear in June.

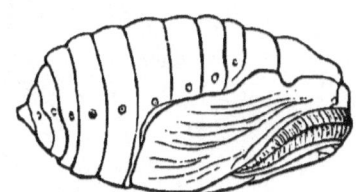

Fig. 280.—Chrysalis of Cecropia Moth. Cocoon reduced.

The Cecropia is dusky brown, and near the middle of each wing is a dull red spot with a white center and a narrow black edging, and beyond the spot a dull red band bordered on the inside with white, and near the tips of the fore wings is an eye-like black spot. The caterpillar is light green, with red and yellow warts covered with short bristles. The cocoon is three inches long, and is fastened to the side of a twig; the outer coat looks like strong brown paper, and inside of this is loose strong silk surrounding an inner cocoon, which contains the chrysalis.

The Promethea is brown with a wavy whitish line near the middle, and with a wide clay-colored border

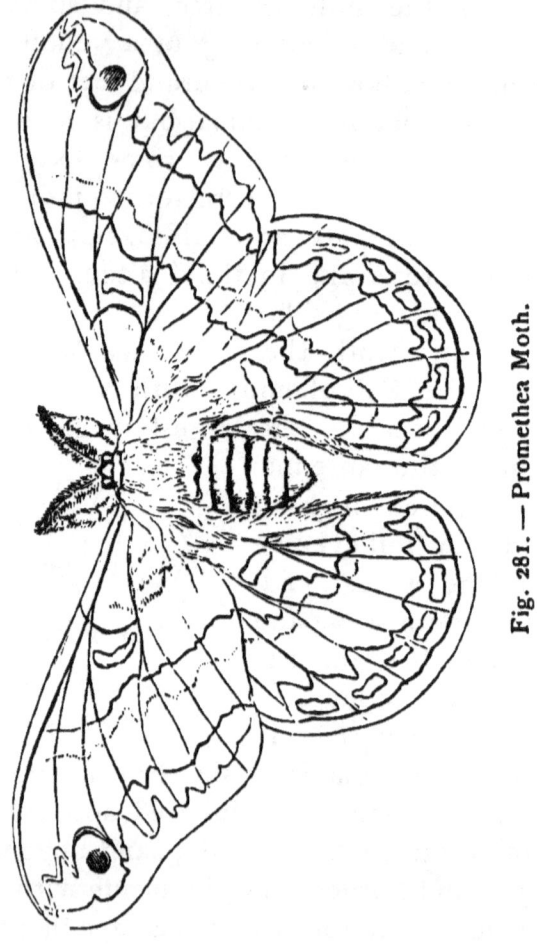

Fig. 281. — Promethea Moth.

marked by a wavy reddish line, and near the tips of the fore wings there is an eye-like spot. The caterpillar feeds upon the sassafras tree. Before making its cocoon, it fastens to the twig, with silken threads, the leaf that is to cover its cocoon, so that it shall not fall in autumn; then it spins its cocoon on the leaf, bending over the edges to cover it.

The Luna, or Pale Empress of the Night, is of a delicate light green color; the hind wings are prolonged into a tail, and each wing has an eye-spot, which is transparent in the center and surrounded by rings of white, red, yellow, and black. The caterpillar lives on the walnut and hickory, and is bluish green, with a yellow stripe on each side, and yellow stripes across the body. It draws together two or three leaves and spins its cocoon inside of them. The cocoon falls with the leaves in autumn, and the next June the beautiful Luna appears.

The Polyphemus Moth is reddish yellow, with a transparent eye-spot, divided by a slender line and encircled by yellow and black, on each wing; on the hind wings adjoining the eye-spot is a large blue spot shading into black.

The American Tent Caterpillar Moth expands an

 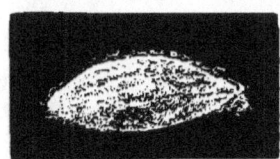

Fig. 282.—Tent Caterpillar Moth.   Fig. 283.— Cocoon of Fig. 282.

inch and a half, and is reddish brown; the fore wings are crossed by two oblique whitish lines. The caterpillars abound in neglected orchards and upon wild cherry trees. The eggs from which they hatch are placed in a cluster on the smaller branches, and covered with a waterproof varnish. They hatch about the time the leaves unfold. The little caterpillars immediately form a small tent between the forks of the branches. As they grow, they enlarge the tent, surrounding it with new layers. They feed at stated times, and return to their tents when they have finished eating. In crawling from one twig to another they spin a silken thread, to guide them back. They rest in their tents at noon and in stormy weather. When full-grown, about the middle of June, they leave the trees, separate, wander about for a time, and at length, in a sheltered place, spin their oval and loosely woven cocoons. The meshes are filled with a thin paste, which becomes a yellow powder. They remain chrysalides about fifteen days.

Geometers, or Spanworms.

The Geometers are Moths whose caterpillars seem to measure the surfaces over which they pass. They are

Fig. 284. — Geometer, or Spanworm.

obliged to move in this way, because they usually have only ten legs, six true legs on the fore part of the body, and four prop legs at the hind extremity. Geometers live upon trees, and let themselves down to the ground by a silken thread which they spin from the mouth while descending. When not eating, many of them stand on the hind legs, with the body extended, and in this attitude may easily be mistaken for a twig. Often when disturbed, they let themselves down, hang till the danger is past, and then climb up by the same thread.

The Cankerworm Moth expands about an inch and a quarter, and the wings are large, thin, and silky. The females have no wings. The larvæ, called Cankerworms, the most destructive of insects, make their appearance about the time the leaves of the apple tree begin to start from the bud. They hatch from clusters of eggs which have been placed upon the fruit and shade trees at various times in and since the autumn before. They immediately commence to eat. They first pierce the leaves with small holes, but as they grow they enlarge these holes, and by and by little more is left than the midrib and veins. When not eating, they lie stretched at full length beneath the leaves. When about four weeks old they reach their full size,—about an inch. They now quit eating, descend to the ground, and, entering to the depth of a few inches, each makes a little cavity, and soon passes into the chrysalis state. Here they remain till after the first frosts of autumn, when they begin to come forth, mainly in the night, in the moth state, and continue to do so, whenever the weather is mild enough, throughout the remainder of the autumn and the winter. They rise in the greatest numbers, however, in the spring. The females crawl up the nearest trees, are joined by the males, and soon begin to lay eggs in rows, forming clusters of sixty to a hundred or more, each cluster being the product of a single female.

Leaf Rollers.

The Leaf Rollers are Moths which, in the caterpillar state, roll up the edges of leaves, fastening them with threads of silk and leaving the ends of the roll open. The moths are small, with the fore wings prettily banded, and sometimes adorned with golden spots.

Fig. 285.—Leaf Roller.

Tineans.

These Moths, in the larva state, gnaw winding passages in the substances upon which they feed. They devour some of the fragments, and fasten together others with silken threads, thus making a covering for their tender bodies. They are the smallest of the Lepidoptera, and are generally very beautiful. They enter through the cracks into closets, drawers, and chests, they get under the edges of carpets, and into the folds of curtains and garments, and here deposit their eggs. In about fifteen days the eggs hatch, and the larvæ immediately begin to gnaw whatever is within reach, covering themselves with the fragments, shaping them into hollow rolls, and lining them with silk. They generally live in these through the summer, become torpid in autumn, change to chrysalides in spring, and in twenty days come forth moths.

Fig. 286.—Tinean.

Two-winged Insects, or Diptera.

Flies, Mosquitoes, the Hessian Fly, Bee Flies, Horseflies, and all their numerous relatives, have only two wings, the place of the hind wings being occupied by two small knobbed threads, called balancers. Mosquitoes have a long bill composed of bristles sharper than the sharpest needles, with which they pierce the flesh of men and animals, and secure the blood upon which they so much delight to feed. The female lays her eggs on the surface of the water, and the larvæ may be seen in great numbers, throughout the summer, in all stagnant pools. They are very lively, and move with a wriggling motion. They rest with the head downward, and with the hind extremity of the body — through which they breathe — at the surface of the water. At length they shed their skins and enter upon the pupa state, in which they live at the surface

Fig. 287. — Horsefly.

Fig. 288. — Bee Fly.

Fig. 289. — Asilus Fly.

Fig. 290. — Horse Botfly.

of the water, and breathe through two tubes on the thorax. In a few days the skin splits on the back, the winged insect appears, and, after resting awhile on its empty skin as it floats upon the water, spreads its wings, and, flies away in search of a victim.

Hessian Fly and Wheat Fly.

The Hessian Fly expands about one fourth of an inch, and has the head, antennæ, and thorax black, the wings blackish and fringed with short hairs. The hind body is tawny, with black on each ring; the legs are brownish, and feet black. Two broods appear in a year, — one in spring and one in autumn. The females lay their eggs on the young blades of wheat, both in spring and fall. The eggs are only about one fiftieth of an inch in length, pale red, and they hatch in about four days, producing pale red maggots. The larvæ immediately crawl down the leaf till they come to a joint. Here they rest a little below the surface of the ground till they have undergone their transformations. They injure the plant by sucking its sap. The larvæ reach their growth in five or six weeks, and are then covered with a hardening, brown or chestnut-colored skin, and the insect is then said to be in the flaxseed state, from its resemblance to a flaxseed. In April and May they complete their transformations, come forth in the winged state, and soon begin to lay their eggs upon the spring wheat, and upon that sown the autumn before. The maggots hatched from these eggs pass down the stem as before stated, take the flaxseed form in June or July, and in autumn most of them are transformed into winged insects; others remain in the ground through the winter, and are transformed in the spring, as before stated. These flies sometimes move in immense swarms in search of fields of their favorite grain where they may lay their eggs. The Hessian Fly received its name from the incorrect belief that it was brought to this country in straw by Hessian troops at the time of the Revolutionary War.

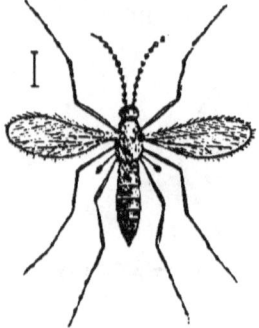

Fig. 291. — Hessian Fly.

The American Wheat Fly is about one tenth of an inch long, orange-colored; wings transparent, eyes black and prominent; antennæ long and blackish, those of the male being twice as long as the body, and consisting of twenty-four joints, and those of the female about as long as the body, and consisting of twelve joints. The Wheat Flies, in their perfect form, appear between the first of June and the last of August. They often move in immense swarms, taking wing in the morning and evening, and in cloudy weather, at which times they lay their eggs in the opening flowers of the grain,— barley, rye, and oats, as well as wheat. The eggs hatch in about eight days, producing little yellow maggots, which are found within the chaffy scales of the grain. The eggs are laid at different times, so that all do not come to maturity together; but they appear to come to their growth in about fourteen days. They prey upon wheat in blossom and in the milk, and do not touch the kernel after it has become hard. At length they cease eating, and soon after shed their skins, after which they become active again, and in a few days descend to the ground. Here they burrow, remain through the winter as larvæ, become pupæ in early summer, and in a few days assume the winged state.

Horseflies.

These are generally large Flies, having a proboscis inclosing very sharp lancets, with which they readily pierce the skin of horses and cattle, in order to suck their blood. They have very large eyes, occupying nearly the whole head. There are several species, and some of the largest are nearly an inch long. The larvæ live in the ground. Figure 287.

Asilus Flies.

These are very long-bodied Flies, and are covered with stiff hairs. They are very rapacious, seizing and bearing away other insects. In the larva state they live in the roots of plants. One kind feeds upon the roots of the pieplant, or rhubarb, of the gardens. Figure 289 shows a common kind of Asilus.

Bee Flies.

These Flies are so named from their general resemblance to Bees. They have a very long proboscis. They frequent sunny places in the woods, in the spring, and fly swiftly, but stop every little while and balance themselves in one place in the air.

Botflies.

These Flies, in the larva state, live in various parts of the body of the ox, horse, and sheep, and occasion great suffering, and sometimes death, to these useful animals. One kind of Botfly lays its eggs upon the fore legs of the horse, another upon the lips, another upon the neck; by biting the parts, the horse swallows the eggs, and the young hatch and cling to the walls of the stomach. The Ox Botfly lays its eggs on the back of cattle, and the larvæ live in burrows in the skin. The Sheep Botfly lays its eggs in the nostrils of the sheep and the larvæ crawl into the head, and often cause death. Figure 290 is the Horse Botfly.

BEETLES, OR COLEOPTERA.

Beetles are Insects whose forward or upper wings are hard and horn-like, and meet in a straight line along

BEETLES.

the top of the back; and there is generally a little triangular piece between the bases of the wings, called the scutellum. The hind, or under wings, are thin, and when the insect is not flying are folded and concealed by the horn-like upper wings. The colors of Beetles are often exceedingly beautiful and brilliant, rivaling even those of precious stones and of birds.

Beetles have two pairs of jaws, which move sidewise, by means of which they bite their food, which in some cases consists of other insects, in others of leaves or other parts of plants. In the larva state, Beetles are called grubs. The kinds are very numerous, probably not less than a hundred thousand in all.

Tiger Beetles.

These are very common in warm sandy places, and may be seen in the roads in the country every pleas-

Fig. 292. — Common Tiger Beetle.

Fig. 293. — Larva of Tiger Beetle.

Fig. 294. — Hairy-necked Tiger Beetle.

ant day. They are very beautifully and often splendidly colored, and have a large head, large eyes, and toothed jaws. They run rapidly, and fly when approached, but soon alight again. They devour great numbers of other insects for food, thus benefiting the farmer and gardener. The larvæ or grubs, are soft, white, and furnished with jaws like the adults; like the latter, they feed on other insects, which they secure by digging holes in the ground, in which they remain, the head just closing the opening of the hole; when some insect comes near enough, they seize it, draw it into the hole, and devour it.

Ground Beetles, or Carabidæ.

These also prey upon other insects, and the kinds are very numerous. They have the jaws very long and hooked, and very long legs. Some of them have no under wings. One kind is called the Caterpillar

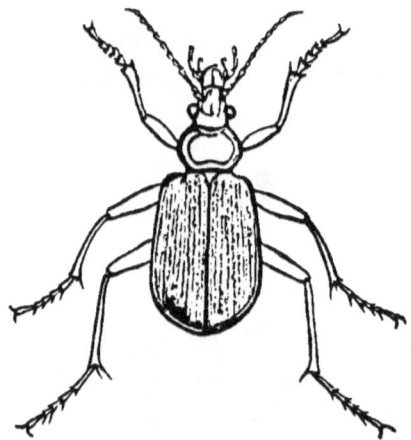

Fig. 295. — Caterpillar Hunter.

Hunter, because it destroys so many larvæ of other insects. It eats great numbers of the Cankerworm, the most destructive insect which has appeared upon our beautiful fruit and shade trees, and which is described on page 166. It appears about the time the Cankerworms leave the trees and come to the ground. The Glowing Caterpillar Hunter is a smaller kind, and is black, with six rows of sunken, brilliant red spots.

Water Beetles.

These Beetles live in the water, and their long hind legs are well fitted for swimming, being fringed on their inner side. They are very voracious, and devour other insects, and, in some cases, young fishes. Some of the species are more than an inch long. The Whirligig Beetles which are found on the surface of still waters, where they look like brilliant spots gliding in all sorts of curves, are much smaller, and belong to another family.

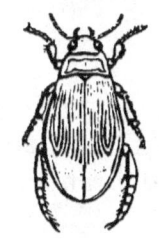

Fig. 296.— Water Beetle.

Carrion Beetles.

Carrion Beetles live together in great numbers in the bodies of decaying animals. Some kinds have the habit of burying the small animals which they find dead, and it is remarkable how quickly they find out where such animals are. If a dead frog, or mouse, or bird is placed upon the ground, these beetles will be seen about it in a few hours; and beginning to dig beneath it, they soon sink it out of sight. The females then lay their eggs in it, so that when the young hatch they find themselves amidst a supply of suitable food.

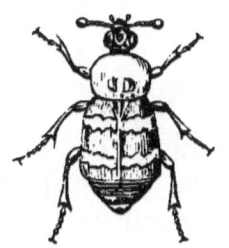

Fig. 297. — Carrion Beetle.

Rove Beetles.

These are long and narrow, with stout jaws, and the hind body much longer than the wing-covers. When they run they raise the hind body and move it in different directions. They are found about decaying substances. The larvæ closely resemble the perfect insect.

Fig. 298. — Rove Beetle.

Horn Bugs.

Horn Bugs are Beetles which have the body very hard and oblong, the thorax and head very large, and the upper jaws large and often curved and branched.

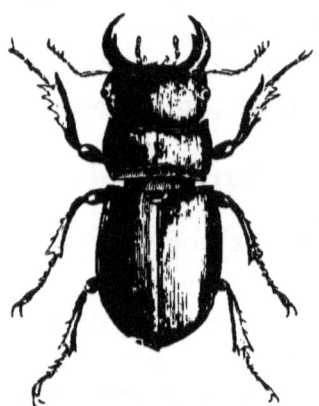

Fig. 299. — Horn Bug.

They keep in hiding in the daytime, and fly about at night. In the adult state they eat the leaves of trees; but the grubs live in the trunks and roots of trees, sometimes for six years before they become Beetles.

Scarabæidæ

The Beetles known as Scarabæidæ have the antennæ ending in a knob made up of three or more leaf-shaped pieces, a sort of plate which extends forward over the face like the visor of a boy's cap, and their legs toothed on the outer sides, thus fitted for digging. Some live on the ground and are called Ground Beetles; others live upon trees, whose leaves they eat, and are called Tree Beetles; others feed upon the sweets of flowers, and are called Flower Beetles. Some kinds are very large, as the Hercules Beetles of South America, which are five

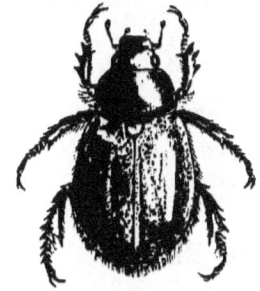

Fig. 300. — Goldsmith Beetle.

Fig. 301. — Phanæus.

inches long. Many are brilliantly colored, and the Phanæus has a horn-like projection on the head. The May Beetles are brown-colored Scarabæidæ, which, attracted by the light, fly into our rooms in the early part of summer; in the grub state they live in the ground, and are white, with a brownish head. The Goldsmith Beetle is of a beautiful golden color above, and copper color, with whitish wool, below. It feeds upon leaves, among which it hides by day, flying in the morning and evening twilight. The Spotted Pelidnota is found on the grapevine in July and August. It is about an inch long, brownish-yellow above, with three black dots on each wing-cover, and one on each side of the thorax.

Many of these Beetles not only injure the foliage of

shrubs and trees, but in their grub, or larva, state they devour the roots of grasses and other plants, and thus do immense injury to the crops. Fortunately, the crow and many other animals devour them eagerly.

Buprestidans.

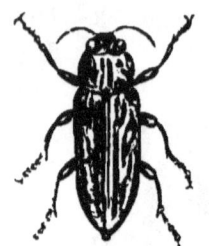

Fig. 302. — Buprestis.

These Beetles, in the larva state, live in the trunks of trees, eating holes in all directions, and injure the trees. Different kinds bore the peach, plum, oak, and pine. The perfect Beetles are long and very solid, with a sunken head, and often with metallic colors.

Spring or Snap Beetles, or Elaters.

When placed upon the back, these Beetles at once, with a snap and a jerk, throw themselves upwards; and they repeat the operation till they come down right side up. They perform this feat by means of a spine-like organ situated on the under side of the breast. Spring Beetles vary from half an inch to two inches in length, and the head is almost concealed in the thorax. One of the most curious kinds has two eye-like spots on the thorax, as seen in Figure 303.

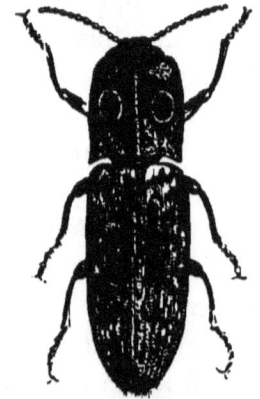

Fig. 303. — Eyed Spring Beetle.

Some of the Elaters, and others closely related to them, give out a brilliant light at night, and are known as Fireflies. They are common in meadows in summer. Some of the tropical kinds emit such a brilliant phosphorescence, that a few of them placed in a glass vessel give light enough for a person to read by.

Curculios, or Weevils.

These Beetles are hard, generally rather small, some being minute, and in most cases they have a long, slender snout. In some, however, the fore part of the head is broad. They feign death when disturbed, and,

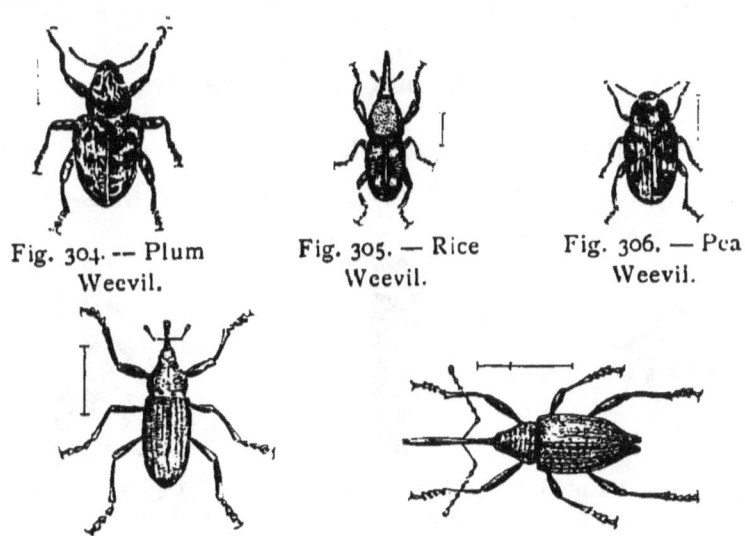

Fig. 304. — Plum Weevil.   Fig. 305. — Rice Weevil.   Fig. 306. — Pea Weevil.

Fig. 307. — White Pine Weevil.   Fig. 308. — Long-snouted Weevil.

if upon a tree, fall to the ground and remain motionless till all is quiet. The Pea Weevil lays its eggs on the pea blossoms, and the grub enters the pea through the green pod, and remains there till the next spring, when it comes out as a perfect beetle or weevil. The Baltimore Oriole splits open the pods for the sake of obtaining the grubs contained in the peas. The White

Pine Weevil, in the larva state, lives in the trunk of the pine, in which it cuts passages in various directions. The Long-snouted Nut Weevil, in the larva state, lives in nuts. The Plum Weevil, when shaken from the tree, looks like a dried bud. This weevil makes a crescent-shaped wound on the surface of the plum, in which it lays an egg; from the egg there hatches a whitish grub, which burrows into the plum, even to the stone. The Rice Weevil feeds upon rice, wheat, and Indian corn. It is about one tenth of an inch long, with two red spots on each wing-cover.

Long-horned, or Capricorn, Beetles.

These Beetles have very long and generally curved antennæ. When caught they make a squeaking sound,

Fig. 309. — Painted Clytus.

Fig. 310. — Larva of Apple Borer.

Fig. 311. — Apple Borer.

by rubbing together the joints of the thorax and hind body. In the larva state they live in the trunks of trees and in timber, and are called *borers*. As they eat their way into the timber they fill the passages behind them with their cuttings. Some, however, as the Apple Borer, keep the ends of their burrows open, out of which they cast their chips. They remain in the larva state from one to three years.

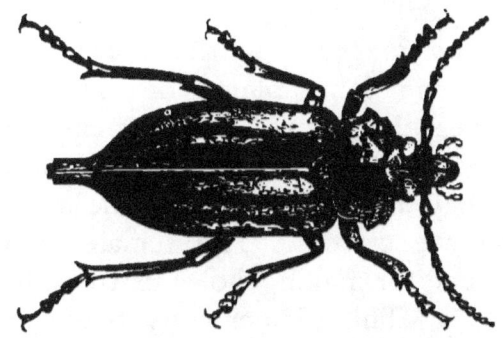

Fig. 312. — Broad-necked Prionus — a Capricorn Beetle.

Chrysomela and Ladybirds.

These are either egg-shaped or hemispherical, and are remarkable for their beautiful colors. The first are

Fig. 313. — Ladder Chrysomelan.   Fig. 314. — Cucumber Beetle — a Chrysomelan.   Fig. 315. — Ladybird.

blue, green, or golden; the latter are black, red, or yellow, with dark spots. The Ladybirds devour plant lice, and are thus of great benefit to the gardener.

BUGS, CICADAS, AND TREE HOPPERS, OR HEMIPTERA.

These Insects have a slender, horny beak, which, when not in use, lies upon the breast under the body.

Cicadas, or Harvest Flies.

The Cicadas, or Harvest Flies, have a very large head, large eyes, three minute eyes on the top of the head, and their wings are large, thin, and very distinctly veined. The males make a very loud buzzing sound by means of curious organs resembling kettledrums, one being placed on each side of the hind body, near the thorax. The ancient Greeks loved to hear the buzzing of the cicadas, and kept them in cages that they might enjoy their rude music. These people also ate cicadas. The females have a very curious piercer for making holes in trees, in which to lay their eggs. This piercer consists of three pieces, the two outer ones grooved on the inside and toothed on the outside like a saw, and a central borer which plays in the groove formed by the outer two.

Fig. 316. — Seventeen Year Cicada.

The Seventeen Year Cicada is about an inch long, the general color black, with the eyes, larger veins, and

Fig. 317. — Dogday Cicada, or Harvest Fly.

forward edges of the wings red. This is generally

called the Seventeen Year *Locust*, but it is in no sense a Locust, and should not be called by this name. The name "Seventeen Year" is given to it from the belief that it appears in the same place only once in seventeen years.

The Dogday Harvest Fly is over an inch long, the body black above, marked with green, and the under side covered with a white substance resembling flour. It appears at the beginning of the dog days, and its singing may be heard among the trees through the middle of the day. The pupæ of this species and of the Seventeen Year Cicada, as they come out of the ground and crawl up the trees, look like Beetles. Soon the pupa skin splits on the top of the back, and from the opening thus made the perfect insect comes forth, leaving the brown pupa skin attached firmly to the tree, and at a little distance looking as when alive.

Tree Hoppers.

These Insects are remarkable for their curious and often grotesque shapes. They live on the sap of trees

 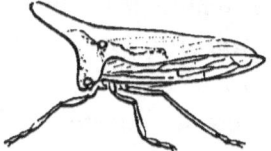

Fig. 318.— Tree Hopper.   Fig. 319. — Same enlarged.

and herbs, and imbibe it in such quantities that it oozes out of the body, often concealing the insect in a mass of frothy matter or foam. Figure 318 shows a common kind, as seen when looking upon its back; Figure 319 is the same in profile, considerably enlarged.

Plant Lice, or Aphides.

These Insects have the body short, and at the hind extremity there are two little tubes, from which come minute drops of a very sweet fluid. Aphides inhabit all kinds of plants, the leaves and softer portions being often completely covered with them. The young are hatched in the spring, and soon come to maturity, and, what is remarkable, the whole brood consists of wingless females; and what is still more remarkable, these females bring forth living young, each female producing fifteen or twenty in a day. These young are also wingless females, and at maturity bring forth living young, which are also all wingless females, and in their turn bring forth living young; and in this way brood after brood is produced, even to the fourteenth generation, in a single season. But the last brood in autumn contains both males and females, which stock the plants with eggs, and then perish. Réaumur, a celebrated naturalist, has proved that a single Aphis, in five generations, may have about six thousand millions of descendants! Wherever Plant Lice abound, ants collect to feed upon the honey-like fluid produced by them; and the most friendly relations exist between these two kinds of insects. An Aphis has been known to give in succession a drop of the fluid to each of a number of ants.

Fig. 320.—Aphis.

Scorpion Bugs.

These Bugs live in the water, and can sting severely. They devour other insects, which they seize with their fore legs, which act as pincers.

Fig. 321. — Scorpion Bug, or Nepa.

Fig. 322. — Squash Bug.

Squash Bugs.

The Squash Bug passes the winter in a torpid state, and when the leaves of the squash appear it lays its eggs in clusters on the under side of them.

Straight-winged Insects, or Orthoptera.

These Insects have wings which lie straight along the top or sides of the back. They do not pass through the larva and pupa states, but the young are constantly active, feeding and growing, and differ from the adults in size, in having only rudiments of wings, and in frequently changing their skins. They shed their skins six times, and then come forth perfect insects.

Earwigs.

These Insects have a pair of sharp-pointed nippers at the hind part of the body, which they can open and

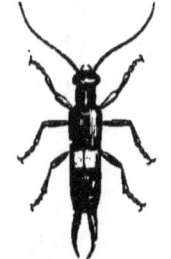
Fig. 323. — Earwig.

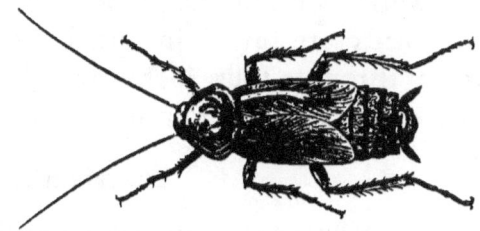
Fig. 324. — Cockroach.

shut like a pair of scissors. They are found under stones and under the bark of old trees, and fly only at night. They are believed by some, probably without reason, to crawl into people's ears.

Cockroaches.

Cockroaches are found in forests, and some species infest kitchens, storerooms, and closets, devouring all kinds of food, and even clothes. Figure 324 shows a kind common here, although it originated in Asia.

Walking Sticks and Walking Leaves.

The Walking Sticks are Insects which look like dry twigs; and the Walking Leaves have wings that look almost precisely like leaves. They belong mostly to

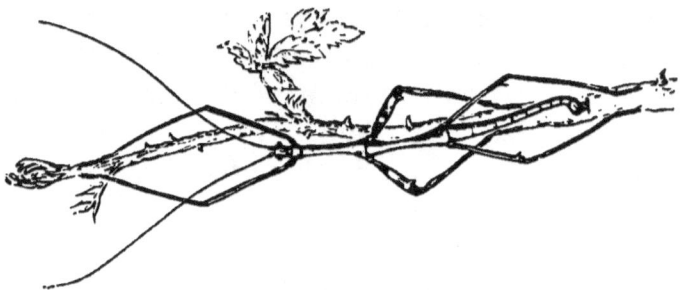

Fig 325. — Walking Stick.

the warm parts of the globe, but several kinds of Walking Stick are found in the United States. They are from three or four inches to a foot long. Figure 325 shows one of our common species, about half size.

The Mantis.

The Mantis is a grasshopper-like Insect which has the fore legs suited for seizing and holding prey. It is found upon plants and trees, where it sits for hours, holding up its fore legs, ready to seize any insect which

Fig. 326. — American Mantis.

comes within reach. Some of the superstitious inhabitants of the East believe that at such times the Mantis is engaged in religious devotions. Figure 326 represents the only kind found in the United States.

Crickets.

Crickets have a flattened body, long antennæ, and long appendages behind. The males chirrup to attract their mates, and this familiar sound is often heard throughout the night. It is produced by rubbing the wings against one another. The most common Crickets of the fields are dark-colored, but some, like the Climbing Crickets, are white. The Mole Crickets have fore feet resembling those of the Mole, and well adapted for digging. They burrow in the ground,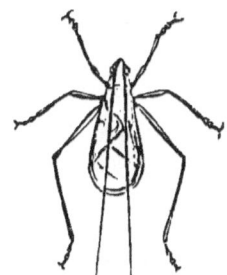

Fig. 327. — White Climbing Cricket.

and prey upon other insects. Some kinds of Crickets take up their abode in houses, and the sound of "the cricket on the hearth" is a familiar one to people who live in the country.

Fig. 328. — Mole Cricket.

Locusts.

These are grasshopper-like Insects which have very long antennæ, a long ovipositor, and many of them

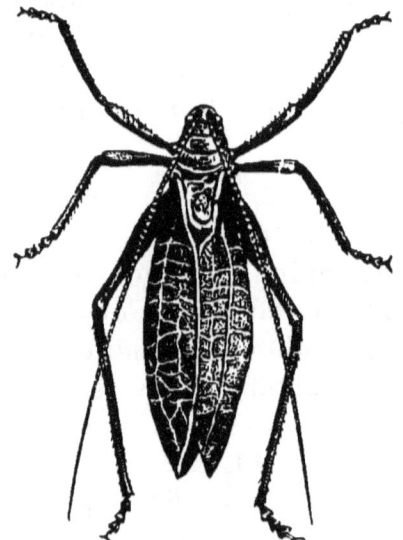

Fig. 329. — Katydid.

produce a grating noise by rubbing their wing-covers together. They are known as Katydids, Leaf-winged Grasshoppers, Sword Bearers, and Meadow Grasshoppers. None are more interesting than the Katydids, whose curious notes are heard at early twilight or on moonlight evenings, and in cloudy days, throughout the autumn. These insects are about an inch and a half long, and the wings shut around the body like the two valves of a peapod. They produce sounds resembling the words "Katy did." A thin membrane is stretched in a strong frame, in the over-lapping part of each wing-cover. The rubbing together of these frames as the wings are opened and shut, makes the sounds.

Migratory Locusts.

These are grasshopper-like Insects having short antennæ, and no long ovipositor. The kinds are many; some tropical ones are three or four inches long. The most common grasshoppers of the United States belong

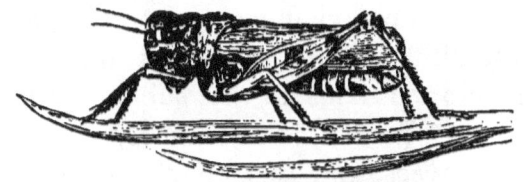

Fig. 330. — Clouded Locust.

in this group, and are familiarly known as the Red-legged Locust, Carolina Locust, Coral-winged Locust, Yellow-winged Locust, and Clouded Locust.

Pseudoneuroptera.

The following insects have netted-wings and were formerly placed with the Neuroptera, or Lacewings. In the larva state, they live in the water.

May Flies, or Ephemera.

Though these insects live only for a few hours or a day in the perfect state, their existence in the larva and semi-pupa states extends through two or three years, and all this time they live in the water. When

Fig. 331.—Stone Fly, half natural size.

Fig. 332. — May Fly.

ready for their final changes, the pupæ crawl to the surface, cast off the pupa skin, and appear at first to be fully developed; this is the sub-imago state; they then fly with difficulty to the shore, affix themselves to plants and trees, and cast off a very delicate covering. After this the wings are brighter, and the tails longer. May Flies appear in such immense swarms in some parts of Europe, that the people collect their dead bodies into heaps to enrich the land. They are common in this country. One of our species is shown in Figure 332.

Dragon Flies, or Darning Needles.

These Insects have a long body, large, lustrous, gauze-like wings, a large head, and very large eyes. They at once arrest our attention by their large size, light and graceful form, variegated colors, and the great velocity with which they speed their way over fields and meadows, or skim the surface of the pools or ponds in search of flies, mosquitoes, and other insects, upon

Fig. 333. — Dragon Fly.

which they feed. In the larva and pupa states they live in the water, and are rather long, broad, and flat, with long, sprawling legs, and they crawl about, or propel themselves by ejecting water from a cavity situated at the hind part of their body. They are very voracious, devouring other insects and even one another. When the time comes for the last change, they crawl up the stems of plants, and, having withdrawn from the pupa skin which remains clinging to the plant, and dried themselves, they spread their wings and dart swiftly away. Though they bite fiercely with their jaws, they have no sting, and are harmless to man.

NET-WINGED INSECTS, OR NEUROPTERA.

Corydalus.

The Horned Corydalus expands five or six inches, and the male has two long, horn-like pincers.

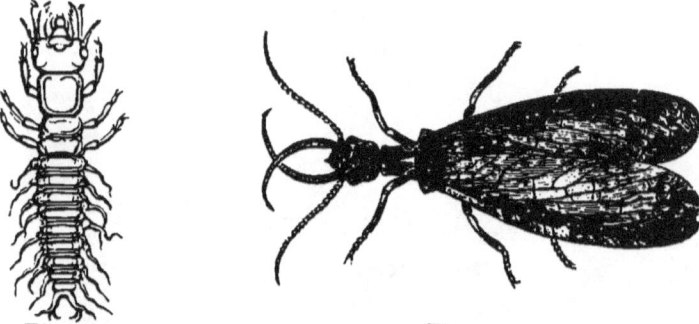

Fig. 334.   Fig, 335.
Larva and perfect insect of Horned Corydalus.   One half natural size.

Lacewings, or Ant-lions.

The Ant-lion is so called because, in the larva state, it preys upon ants and other insects, which it secures in the following manner: it makes a pitfall, or cavity,

Fig. 336. — Ant lion.

Fig. 337. — Larva of Ant-lion.   Fig. 338. — Pitfall of Ant-lion.
Enlarged.

Figure 338, at the bottom of which it conceals itself, excepting its jaws, and there awaits its prey. Whenever an insect falls into the pit, the Ant-lion seizes and devours it.

Caddice Flies.

These Insects are the most interesting while in the larva state. They live at the bottom of ponds and

Fig. 339. — Caddice Fly.

streams, in cases which they construct of bits of wood or grasses, of grains of sand, or of fragments of broken shells, and which are lined with silk spun from their mouths. They sometimes load one side of the case with heavier pieces, in order to keep that side downward.

SPIDERS AND SCORPIONS, OR ARACHNIDS.

Spiders.

Spiders have the body divided into only two well-marked portions, — the head and the hind body. They have eight legs, and two palpi or feelers resembling legs, but no wings, and do not change form from the young to the adult state. Most kinds feed upon insects.

Many Spiders have, at the hind part of their body, a most wonderful organ, called the spinneret, by which the delicate threads of the spider web are spun. It consists of four to six knobs, with a thousand or more holes in each knob. Through these the invisible silken threads pass out, — more than four thousand at a time, — and at a little distance from the knobs all these unite into one, forming the single line of spider web which all are so familiar with. As the threads issue from the knobs they are a sticky fluid, — which has

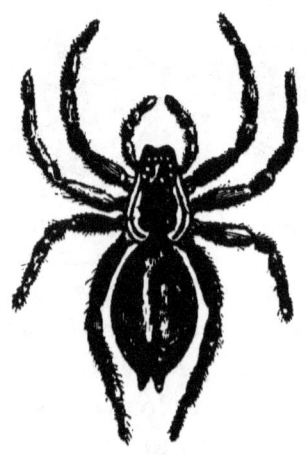

Fig. 340. — Spider — Lycosa.

been secreted in little bags in the abdomen; but this hardens into silk as soon as it comes to the air. The length of the line which the Spider is able to produce is truly wonderful. Dr. Wilder wound nearly two miles of silk, in less than a day, from his celebrated *Nephila plumipes*, — a Spider which he discovered in South Carolina. The kinds of Spider are very numerous, and most of them spin some sort of a net-like web, in or near which they live, and by means of which they capture insects for food. The House Spider spreads a flat net in the corners of rooms. The Geometric Spider spreads a vertical net, which is made in the most beautiful manner, radiating lines running from the center, like the spokes of a wheel, and these connected by a spiral line, which at a little distance gives the appearance of lines arranged in circles from the center outwards. Some kinds of Spider have, near the principal web, a silken retreat, or den, where the owner hides till the quivering spider lines which run into its office telegraph the fact that a fly has become entangled; instantly the spider rushes out of its retreat, pounces upon the victim, and bites it, if possible, putting into the wound a fatal poison. If the insect is too powerful for the spider, the latter waits till the insect gets more entangled, and finally exhausted, by its efforts to escape, then binds it with silken bands, and begins to devour it. The bite of an ordinary spider will kill a fly; the bite of some of the large kinds in South America kills the humming bird. The female spiders lay eggs and inclose them in silken sacs. Some kinds carry the egg-sac about with them; others spin it in a safe place, and, in some instances, stay near to guard it, and to tear open the egg-sac as soon as the young are hatched, that they may escape. One of the most curious of these egg-sacs is that shown in Figure 341,

Fig. 341. — Egg-case of a Spider (the Vase-maker).

and which was made by some Spider which may properly be called the *Vase-maker*. Two "vases," like the one in the woodcut, were found standing about a foot apart on the stem of a grapevine. The outside of the vase looks like brown paper, or it is in appearance and in toughness like the outside of the cocoon of the Cecropia Moth; the vase is fastened to the vine by a vast number of threads of silk passing from one side of the vase to the other around the vine, and the threads are so nicely arranged that the vase cannot turn nor slip from its place. On opening this curious structure, it was found to be filled with the finest silk and a great number of newly-hatched spiders.

Scorpions.

The Scorpions are confined to warm regions, and live among ruins of buildings, under rubbish, and sometimes in houses. They have a long body ending in a

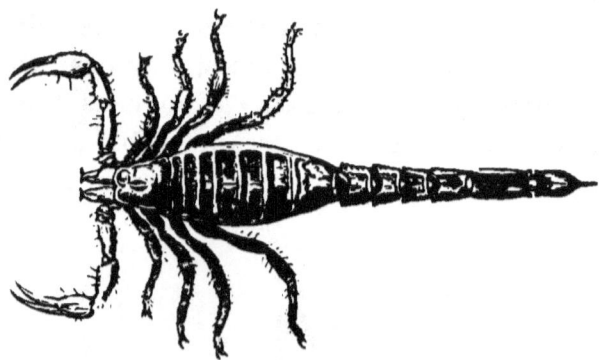

Fig 342. — Scorpion.

curved, sharp sting, with which they inflict painful wounds, which may be dangerous or even sometimes fatal. They can run quite rapidly, and can bend the hind body or tail in any direction, and use it both for attack and defence. The form above is found in Texas.

CENTIPEDES, OR MYRIAPODS.

These are long and worm-like, and divided into very numerous rings or joints; each joint generally bears two pairs of feet. In the temperate zones they are

Fig. 343. — American Myriapod, or Galley Worm.

Fig. 344. — American Earwig, or Lithobius.

only two or three inches long; tropical species are sometimes a foot long, and their bite is often very poisonous.

CRUSTACEANS.

These Arthropods have a crust or shell, the head and thorax often united into one piece; they live in the water and breathe by gills. Some kinds, however, live upon the land. They feed upon all sorts of animal food, and shed and renew their shells many times.

CRABS, LOBSTERS AND SHRIMPS, OR TEN-FOOTED CRUSTACEANS.

Crabs can walk forward, backward, and sidewise. The tail, or hind body, is small, and is doubled under the forward part of the body, where it fits into a groove. The kinds of Crab are very numerous, and some are found on every seacoast. They vary in size from that of a penny to those which, with the legs outspread, cover a space a yard square. Some kinds are very much prized for food; the one shown in Figure 347 is sold in great numbers in the markets of New York and Philadelphia. Hermit Crabs have the hind part of the body long, soft, and tapering, and they take up their abode in empty univalve shells, which

Fig. 345. — Fiddler Crab.

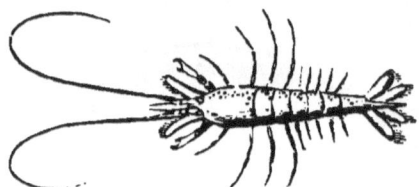

Fig. 346. — Bait Shrimp.

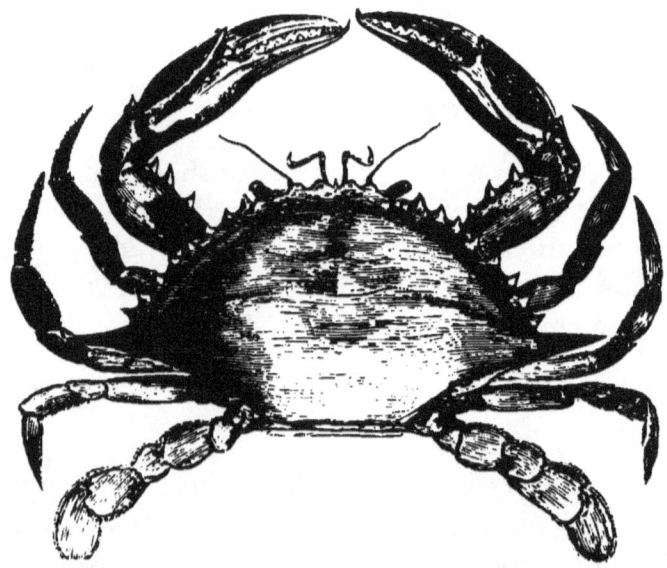

Fig. 347. — American Edible Crab.

they drag about with them wherever they go, and they look as though they were the real and original owners of the houses which they live in. When a Hermit Crab becomes too large for the shell which it has chosen for its home, it abandons it, and begins its search for a new one, inserting itself backwards into one shell after another till one is found which suits it. When not moving about, or when alarmed, it retreats as far as possible into the shell, and closes the opening with its larger claw.

Lobsters and Shrimps have the hind body, often called the tail, large and long, and generally turned forward, as seen in Figure 348. The American Lobster is from one to two feet long, and weighs from three to ten pounds or more. It is very abundant on the coast of New England, and great numbers are caught in lobster pots baited with fish, and are sold in the markets of Boston, New York, and other cities.

Two of the forward leg-like appendages of Lobsters are greatly enlarged, and end in powerful claws or pincers. One of these is provided with blunt teeth, or tubercles, suited for crushing shells, and the other with exceedingly sharp teeth suited for biting. So powerful are these organs that with them a Lobster can easily bite off a man's finger; and if one were to get hold of your hand, you could release it only by breaking off the Lobster's claw. The fisherman, well knowing their biting powers and habits, puts a wooden plug into the joints of their pincers, so that they can not open them; if this were not done, the lobsters, when confined in the lobster car, — a large box in the water where lobsters are kept after they are caught, — would bite off the limbs of one another. In crawling the lobster moves rather slowly, but sometimes, by a single stroke of its powerful tail or hind body, it darts through the water, backwards, a distance of fifteen or twenty feet, with the swiftness of an arrow. When a

Lobster or other Crustacean loses a leg or other organ, another like it grows to supply its place. But one of the most remarkable facts about Lobsters and other Crustaceans is, that from time to time they shed the shell in one piece, so that the cast-off shell looks exactly like the perfect animal,— antennæ, eyes, jaws, legs, paddles, and even every hair, are all just as they were when they covered the live Lobster. The Lobster

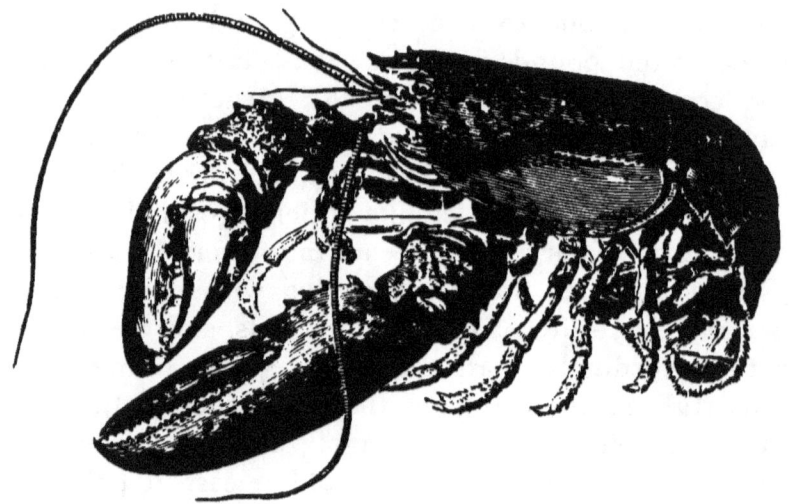

Fig. 348. — American Lobster.

comes out of its shell through a rent on the back, and is at first very soft; it at once increases in size, and in a few days its skin becomes as hard as the shell which it cast off. This shedding of the shell is necessary for the growth of this animal, for after the new shell hardens further growth is impossible. When a Lobster is ready to shed its shell, there are two hard, stone-like bodies at the sides of the stomach, and it is supposed that these furnish a part of the solid matter for the new shell; for they immediately begin to grow smaller after the moulting, and soon entirely disappear.

The Crawfish, or Fresh-water Lobster, much resembles the American Lobster, but is only three or four inches long, and lives in streams and lakes. One kind is common on the western prairies; it lives in holes which it digs in the ground deep enough to find water.

SAND FLEAS, ETC., OR FOURTEEN-FOOTED CRUSTACEANS.

Beach, or Sand, Fleas are little shrimp-like Crustaceans, common on the seabeach. They have seven

Fig. 349. — Sand Flea.

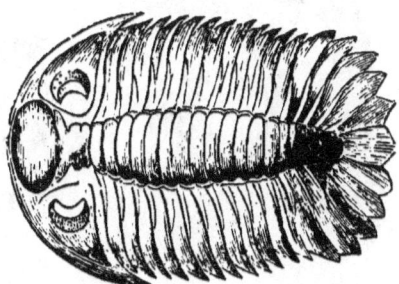

Fig. 350. — Trilobite.

pairs of feet. Imbedded in the rocks are found the curious, closely related Trilobites, which lived ages ago.

BARNACLES AND HORSESHOE CRABS.

The Barnacles are of many kinds. Some resemble bivalve shells, and grow in clusters, attached by stems, as seen in Figure 351; others, as in Figure 352, are acorn-shaped, and are fixed directly upon the rocks, shells, lobsters, or ship-bottoms. They are all provided with feather-like arms or feet, which they regularly protrude and withdraw, — a sort of grasping motion as though they would secure any little animals or particles of food that might be within their reach. Some kinds of Acorn Barnacles completely cover the rocks between high and low water mark; others delight in deep water. In long voyages Barnacles sometimes become so numerous on the bottom of a vessel as to seriously hinder its progress. Although in the adult state Barnacles, or Cirripeds, are fixed and stationary, the young swim freely about.

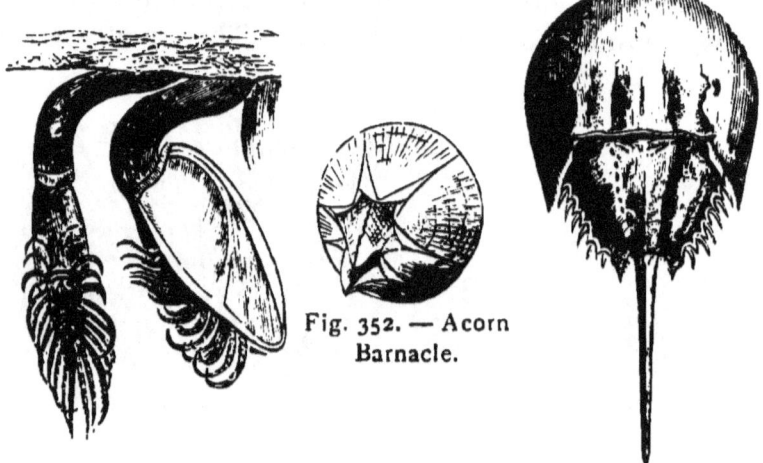

Fig. 352. — Acorn Barnacle.

Fig. 351. — Duck Barnacle.

Fig. 353. — Horseshoe Crab.

The Horseshoe Crab is found on the Atlantic coast of the United States and on the coast of Eastern Asia. Some are two feet in length, and in all cases the body ends in a sharp spine, which some of the savage tribes use for spear points. This curious Crab walks and eats with the same organs, — the lower part of the first six pairs of legs being used for walking, and the upper parts of the same legs being provided with teeth-like organs, and used for jaws.

MOLLUSKS, OR SOFT-BODIED ANIMALS.

The term Mollusk comes from a word which means *soft*, and these animals have a soft body with no backbone nor internal skeleton; nor is the body divided into rings or joints, as in the Arthropods and Worms. Most of them have a hard covering called a shell, and are often called *shellfish;* but they are in no way related to Fishes. The shells are the parts which we oftenest see; for when the animal is dead, the soft parts soon disappear, and only the shell remains. Curious and wonderful as the shells are, they often give only the faintest idea of the appearance of the animals when alive. See the differences

Fig. 354. — Helix. Alive.

between Figures 354 and 355, where the first represents the shell alive and the animal expanded, the second the shell as when dead, or when the animal is concealed in the shell. It is important to know that the shell is a part of the animal and not a mere house which it enters and leaves at pleasure; although it readily expands much beyond the limits of the shell, and with-

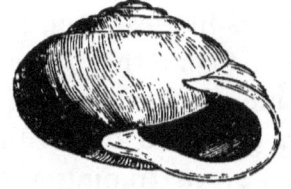

Fig. 355. — Helix. Dead.

draws itself wholly within the same again. Mollusks have, in a greater or less degree, the senses of the higher animals, though they greatly differ among themselves in this respect.

The kinds of Mollusks are very numerous, — not less than fifteen or twenty thousand. They abound in the sea, on the marshes, in pools, streams, ponds, and lakes, and on the land; and they are full of interest when we study them, and all serve some important purpose. They are the food of many other animals. The Right Whale feeds upon small kinds which swim freely in the open sea; the Cod and Haddock and many other useful fishes fatten upon those gathered near or on the bottom; and sea birds feast upon those left bare by the tide. Man reckons the Oyster, Clam, and Scallop among his choicest dishes; and in seasons of scarcity the poor inhabitants on many a seacoast depend upon Mollusks for a large part of their daily food. These animals also furnish the bait for all the extensive fisheries of the north Atlantic. Some of them yield rich dyes. The celebrated Tyrian purple of the ancients was obtained from sea snails.

The shells of Mollusks are of limestone, or carbonate of lime. Pearly within, and of soft and delicate colors, they are often exceedingly beautiful, and are eagerly sought for. The child gathers them for toys, and thinks he hears the roaring of the sea as he puts them to his ear; the savage wears them as ornaments, and some of them as marks of chieftainship; some kinds are gathered by civilized nations and used instead of money in trading with barbarous tribes; other kinds are gathered and wrought by skillful hands into almost numberless articles of use and luxury; and the true naturalist, more enthusiastic than all others, traverses sea and land, and cheerfully endures hunger, thirst, and fatigue, that his collection of shells may lack neither Argonaut nor Nautilus, Cone, Cowrie nor Wentletrap, Helix nor Limnæid, Pecten, Mother-of-pearl nor Unio, nor any other which will enable him to understand more clearly this department of the animal kingdom, and the works of God as revealed in these wonderful objects.

ARGONAUTS, CUTTLEFISHES AND SQUIDS, OR CEPHALOPODS.

These animals all live in the ocean, have a mouth armed with a stout beak, resembling that of a Parrot, a large eye on each side of the head, and surrounding the mouth are long, muscular arms, or tentacles, covered with cup-like suckers, by means of which they cling with the greatest firmness to whatever they lay hold of,—it being easier to tear away an arm than to release it from its hold. They have within the body a sac containing an ink-like fluid, with which they cloud the water, and thus conceal themselves whenever they wish to escape from an enemy. The word Cephalopod means *head-footed*, and is given to these Mollusks because their locomotive organs are attached to the head, as just described. Cephalopods vary from a few inches to many feet in length, according to the kind. They have a most wonderful power of changing their colors, their hues varying constantly. They swim by means of their arms, or with them crawl on the bottom with the head downward. They are very voracious, devouring fishes and other animals, whose flesh they readily tear in pieces with their stout hooked beaks.

Fig. 356. — Argonaut, or Paper Sailor. Much reduced. Warm seas.

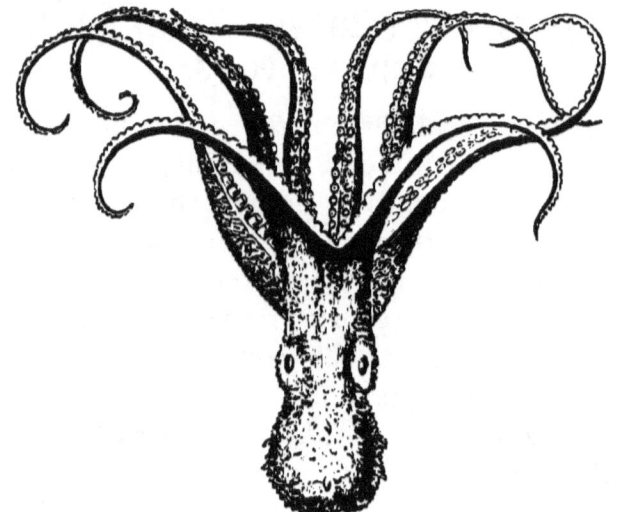

Fig. 357. — Octopus, or Poulp. Much reduced. Mediterranean.

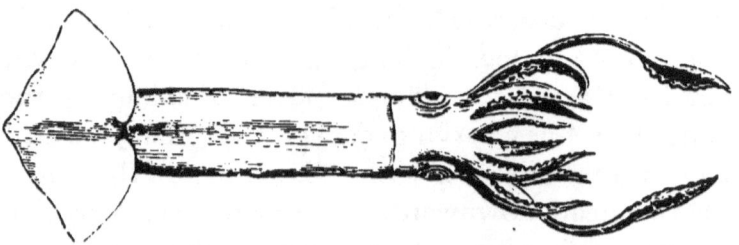

Fig. 358. — Squid, or Loligo. Much reduced. Atlantic Coast of United States.

Cephalopods sometimes reach an enormous size. Aristotle tells us of one which was five fathoms in length, and even larger ones have been seen in recent years. Some have been found with arms thirty feet long and bodies nearly twenty feet. These gigantic Squid have given rise to the tales of the Kraken.

Paper Sailors or Argonauts.

The Argonaut, or Paper Sailor, Figure 356, has a very delicate and beautiful shell, and swims by placing two of its arms, which are webbed, close to the sides of the shell, and the others close together, and then ejecting water from the funnel seen just below the eye. The Argonaut is often called a Nautilus, — the true Nautilus is another animal, — and it has frequently been erroneously stated that it sails on the sea by spreading its sail-shaped arms to the breeze.

Octopus.

The Octopus, or Poulp, Figure 357, has no outside shell, and the arms are united at the base by a web. It varies from one or two inches to two feet in length, and has only eight arms.

Squids, or Loligos, and Cuttlefishes.

Squids have a long body; broad, fin-like organs at the hind extremity; and a long and slender internal shell which, from its shape, is called a "pen." They are from one to two feet and a half long, and, like Cuttlefishes, have ten arms, two of which are longer than the others. By filling their body with water, and then forcibly ejecting it, they send themselves backwards through the water with the swiftness of an arrow. Immense numbers are used for bait in the cod fisheries. The giant squid are related in structure to these smaller ones.

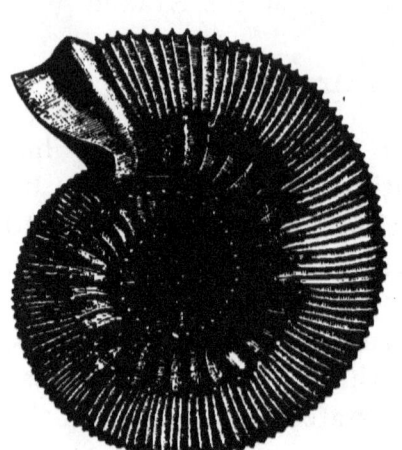

Fig. 359. — Ammonite.

Fig. 360. — Spirula.

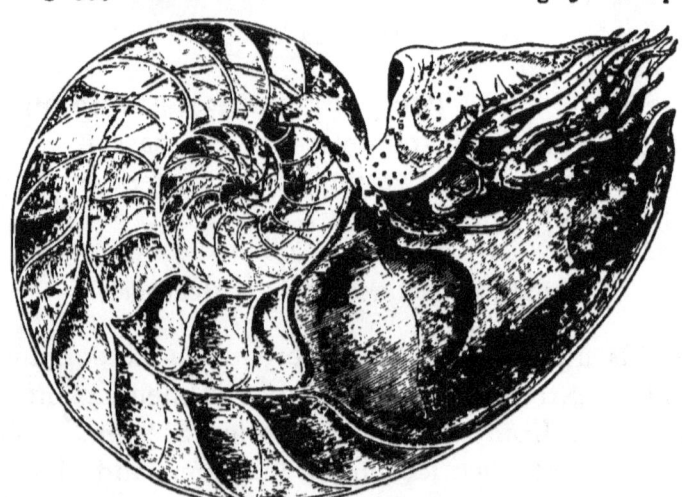

Fig. 361. — Pearly Nautilus. Much reduced. Pacific and Indian Oceans.

Cuttlefishes resemble Squids, but have two of the arms or tentacles much lengthened and expanded at their tips; and they have a broad, internal shell, called *cuttle bone*. This is the cuttle bone which is given to canary birds. On the coasts of the Eastern Mediterranean, Cuttlefishes are so abundant that the cuttle bones are thrown up by the waves into ridges miles in length. Like other Cephalopods, Cuttlefishes have the power of clouding the water by ejecting an inky fluid into it when they wish to escape. This ink, when dried and prepared, is the *sepia* used in painting.

Spirula.

The Spirula resemble those just described, but have a coiled shell inside, Figure 360, and the shell is divided by partitions into chambers.

Nautili and Ammonites.

The Nautilus is the only living Cephalopod which has an external chambered shell. Figure 361 shows the Nautilus as it appears when cut open; the animal lives in the outer chamber, which communicates with all the others by means of a tube called the siphuncle. It has occupied each chamber in turn, making a partition behind as often as it outgrew its old home.

The Nautilus lives in moderately deep water about the islands of the East Indian Archipelago.

Ammonites, Figure 359, are chambered-shelled Cephalopods that lived in the seas ages ago; hundreds of kinds of these, from an inch to a yard in diameter, are found imbedded in the rocks.

SNAILS, OR GASTROPODS.

The term Gastropod means *stomach-footed*, and is given to these animals because the lower side serves them as a sort of foot, by means of which they creep along. But this "foot" is in no way related to the feet of the backboned animals. Most of the Gastropods have a shell; and, as this is made of only one piece, or valve, they are often called Univalves. Some, however, have no shell in the adult state, though all have a shell when first hatched. Most Gastropods have a lid or door, called the *operculum*, with which they close the opening to the shell when they withdraw within. It is a horny plate, sometimes strengthened by shelly matter. Their eyes are two, and often on long stalks, as seen in Figure 354. Many of the

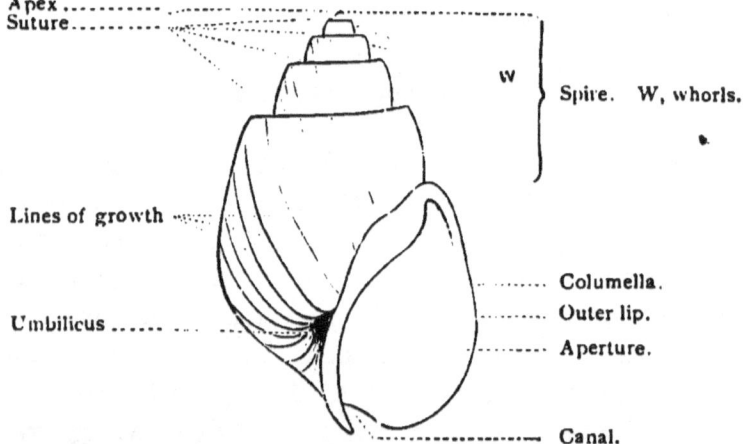

Fig. 362. — Names of the parts of a Gastropod Shell.

Gastropods have horny jaws; but one of the most curious parts of these animals is the tongue, or lingual ribbon, which is a band armed with a great number of glossy silicious teeth, arranged in rows in the most regular manner, and differently in different kinds. The tongue of some kinds contains one hundred and sixty rows of teeth, and one hundred and eighty teeth in each row, or more than twenty-eight thousand in all. The tongue is used like a rasp to scrape the animal's food to pieces.

Many of the Gastropods feed upon vegetable substances, and these have the aperture of the shell entire. The others feed upon animal substances, and have the aperture notched, or drawn out into a canal, as in Figures 363–375. Some of these feed upon dead animals; others attack living mollusks, though shut tightly within their shells, for the Gastropod, with its rasp-like tongue, files a round hole through the shell, then leisurely feasts upon its contents. Thus clams and other large mollusks fall a prey even to some of the very small carnivorous Gastropods, which are among the most dangerous enemies of the oyster, and cause heavy losses every year to the oystermen.

The Gastropods are divided into Air-breathers, as Land Snails and Pond Snails; and Water-breathers, as Sea Snails and River Snails. The young of the former are like the parents, only smaller; the young of the latter differ from their parents, and swim with a pair of fins springing from the sides of the head.

Water-breathing Snails.

Strombs and Conches, or Wing-shells, etc.

These are large marine shells, some of them the largest of the Gastropods. One kind, called the Fountain Shell, is extensively used for making shell cameos;

three hundred thousand of this kind were carried from the West Indies to Liverpool in a single year. The interior of the Conch is of the richest rosy hue.

Murex Shells.

The Murex and its relatives are marine, and prey upon other mollusks. The Spiny Murex of the Moluccas,

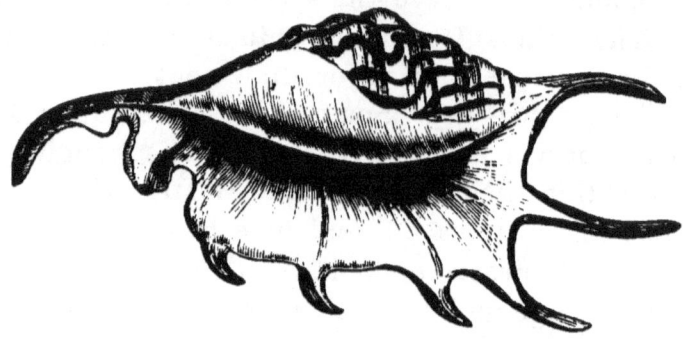

Fig. 363. — Scorpion Shell, or Pteroceras. Much reduced. Chinese Seas.

Fig. 364. — Aporrhais. Coast of New England.

Fig. 365. — Stromb, or Conch. Much reduced. West Indies.

the Pyrula and Tritonium of the coast of the United States, and the Frog Shell of Australia, are some of the principal ones. The ancients obtained the Tyrian purple dye from the Murex Gastropods.

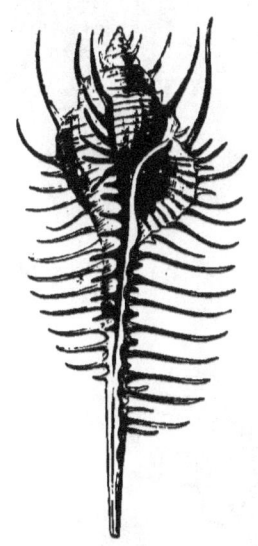

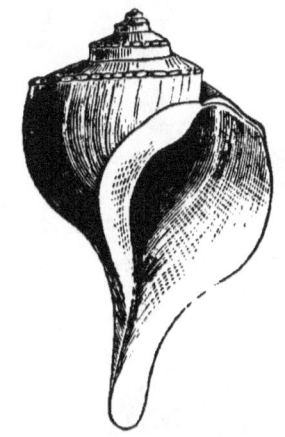

Fig. 367. — Pyrula. Much reduced. Coast of United States.

Fig. 366. — Murex. Much reduced. Moluccas.

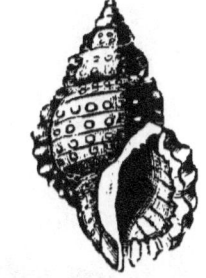

Fig. 368. — Tritonium. Coast of New England.

Fig. 369. — Frog Shell, or Ranella. Reduced. Australia.

Whelks.

The Whelk is one of the most common of the Gastropods. Fig. 370 shows one species as it appears

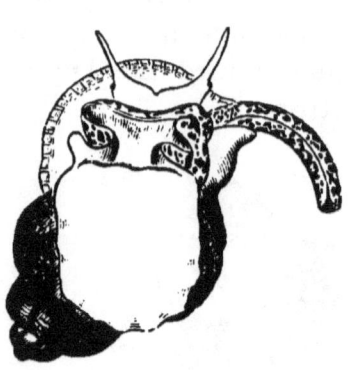

Fig. 370. — Whelk, or Buccinum. North Atlantic.

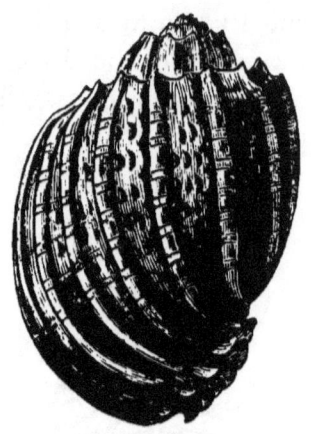

Fig. 371. — Harp Shell. Reduced. Mauritius.

Fig. 372. — Oliva. Reduced. Panama.

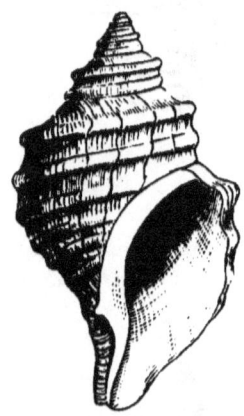

Fig. 373. — Fusus. United States.

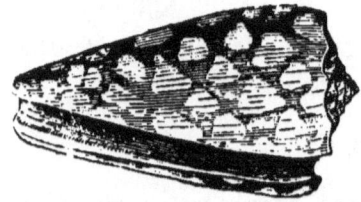

Fig. 374. — Cone Shell. Reduced. China.

Fig. 375. — Ricinula.

when crawling up the glass sides of the aquarium with the foot towards you. The Fusus, of the coast of the United States, may be found upon the shore after storms. The Harp Shell, of the Pacific, is always admired for its beautiful form and its delicate colors. The Oliva Shell, of Panama, is very beautiful, and is taken alive by bait attached to lines.

Cones.

There are nearly a thousand kinds of Cones. They are shaped like a cone with the top downwards.

Volutes.

The Volutes, Miter Shell, and Marginella belong under this head. Figures 376–378.

Cowries.

The Cowries are abundant in the warm seas, and are found on reefs and under rocks. The shell has a shining enameled surface, and many kinds are beautifully spotted and clouded. The Asiatic islanders use them to adorn their clothing, for sinkers to fishing nets, and in trading. The Money Cowrie is brought in immense quantities from the Pacific to England, then carried to western Africa, and used for money in trading with the natives. It is scarcely an inch long. The Egg Cowrie and the Cypræa of the Indian Ocean show the general form of these shells. Figures 379–381.

Naticas, Pyramid Shells, Cerithiums, etc.

The Naticas are Sea Snails which have the shell somewhat globe-shaped. The Pyramid Shells are named

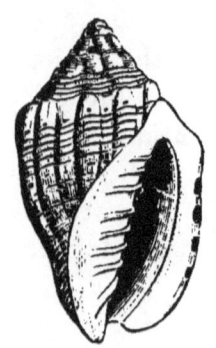

Fig. 377. — Marginella. Reduced. W. Africa.

Fig. 376. — Volute. Much reduced. West Indies.

Fig. 378. — Miter Shell. Much reduced. Ceylon.

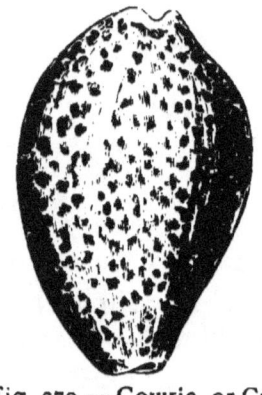

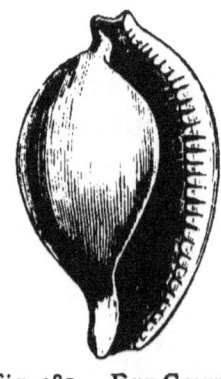

Fig. 380. — Trivia. Great Britain.

Fig. 379. — Cowrie, or Cypræa. Much reduced. Indian Ocean.

Fig. 381. — Egg Cowrie. Much reduced. New Guinea.

Fig. 382. — Sigaretus. West Indies.

Fig. 383. — Natica. New England.

Fig. 384. — Pyramid Shell. Reduced. Great Britain.

Fig. 385. — Pyramid Shell. W. Indies.

WATER-BREATHING SNAILS. 217

  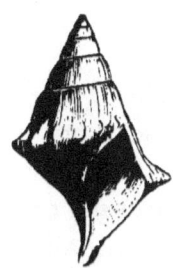

Fig. 386. — Cerithium.   Fig. 387. — Melania.   Fig. 388. — Io.
Much reduced. Moluccas.   Western States.   Southern States.

Fig. 389. — Tower   Fig. 390. — Wentletrap.   Fig. 391. — Worm-
Shell, or Turritella.   Reduced. China.   shell, or Vermetus.
West Indies.   West Indies.

Fig. 392. —   Fig. 393.—   Fig. 394. —
Periwinkle,   Lacuna.   Valvata.   Fig. 395. — River Snail Shell,
or Litorina.   U. States.   or Paludina. United States.

from their shape; the Cerithiums from a word which means a *horn*. The Melanias are fresh-water shells, common in the Western and Southern States.

Wentletraps, etc.

The Tower Shell and the Worm-shell of the West Indies, and the true Wentletraps of the tropical and temperate seas, belong in this group. The Royal Staircase, or Wentletrap, Figure 390, was formerly very valuable, and has been sold for a hundred pounds sterling, although now worth only a few dollars.

Periwinkles.

Periwinkles live in the sea near the shore. Two species are shown in Figures 392 and 393. They feed on algæ, — marine plants.

River Snails.

These live in fresh water, have the shell covered with a green skin, bring forth their young alive, and the embryo snails, even before birth and when so small that they can scarcely be seen without a microscope, have a perfectly formed shell, a "foot" and operculum, delicate tentacles, and distinct black eyes.

Violet Snails.

The Violet Snails live together in large numbers, in the open sea, where they float by means of many air-vessels, which form a raft, *a*, Figure 402. The shell is thin, the base deep violet color, and the spire almost white. They yield a violet dye.

WATER-BREATHING SNAILS.

Fig. 396. — Neritina. Pacific.   Fig. 397. — Nerita. Sinde.   Fig. 398. — Trochus. Great Britain.

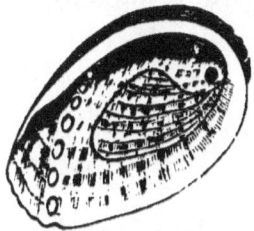

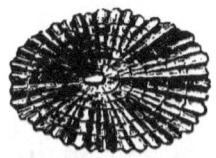

Fig. 399. — Ear-shell, or Haliotis. Reduced. Great Britain.   Fig. 400. — Cup-and-saucer Limpet. Philippines.   Fig. 401. — Keyhole Limpet. West Indies.

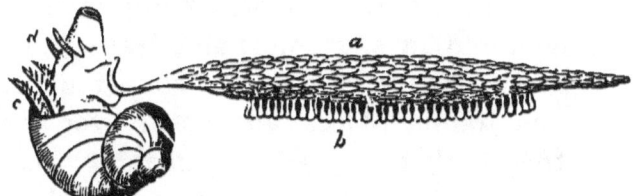
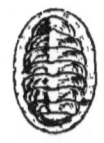

Fig. 402. — Violet Snail. Atlantic.   Fig. 403. Chiton. New England.

a, raft; *b*, egg capsules; *c*, gills; *d*, tentacles and eye-stalks.

Fig. 404. — Rock Limpet, or Patella. New England.

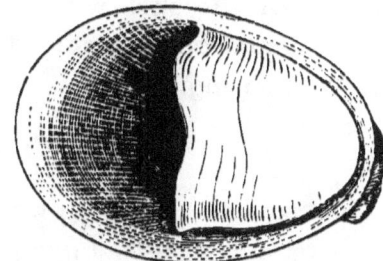

Fig. 405. — Toothshell.   Fig. 406. — Crepidula. New England.

Limpets.

Limpets cling tightly to stones and shells, and move about but little. They are all marine. In England, they are used by fishermen for bait, and on the coast of Berwickshire twelve millions have been collected yearly for this purpose. In the north of Ireland the people collect it for food. On the western coast of South America there is a Limpet a foot across, and the natives use its shell for a basin.

AIR-BREATHING SNAILS.

Land Snails.

Land Snails are very numerous, more than four thousand kinds being already known. Figures 407-412. They all feed upon plants. One of the largest and most common is the *Helix albolabris*, Figure 407, easily found under old logs, stumps, and leaves. In warm, damp weather, Snails of this and similar kinds come out of their hiding places, and crawl over the leaves and up tree-trunks. In early summer they lay eggs in the loose soil beside logs or stones, and in twenty or thirty days the young hatch. When cold weather comes they seek a sheltered spot, close the mouth of the shell with a thin membrane which they secrete, and become torpid, remaining so till the warm days of spring.

Pond Snails, or Limnæidæ.

These live in fresh waters, and lay their eggs in transparent masses on aquatic plants and on stones. They have a thin and horn-like shell. Figures 413-

AIR-BREATHING SNAILS. 221

415. They feed on plants, and glide along the surface of the water, shell downwards. They thrive well

Fig. 407. — Helix albolabris.

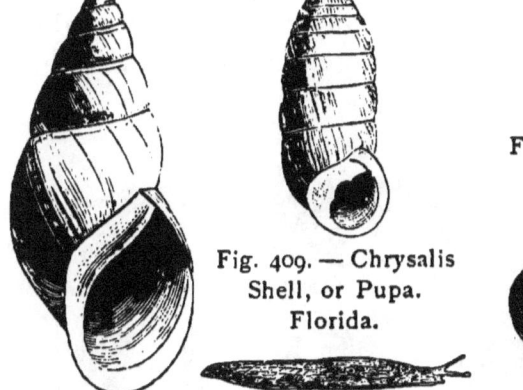

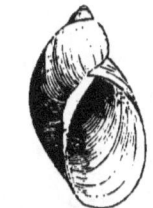

Fig. 411. — Succinea. Western States.

Fig. 409. — Chrysalis Shell, or Pupa. Florida.

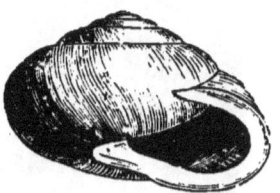

Fig. 408. — Bulimus. California.   Fig. 410. — Slug, or Limax. New England.   Fig. 412. — Helix. Northern States.

Fig. 414. — Planorbis. United States.

Fig. 413. — Physa. United States.

Fig. 415. — Limnæa. United States.

in an aquarium, where they are also very useful, devouring the green confervæ that grow on the glass.

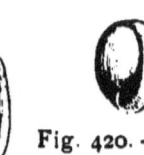

Fig. 416. — Helicina. U. States.

Fig. 417. — Clyclostoma. United States.

Fig. 418. — Acicula. Great Britain.

Fig. 419 — Tornatella. Great Britain.

Fig. 420. — Bulla. U. States.

Sea Slugs.

These have no shells, and many of them only slightly resemble the Gastropods before described. See Figures 421–425.

Fig. 422. — Doris. Great Britain.

Fig. 423. — Elysia. Great Britain.

Fig. 424. — Atlanta. South Atlantic.

Fig. 421. — Eolis. Great Britain.

Fig. 425. — Tritonia. Great Britain.

HETEROPODS AND PTEROPODS.

These live in the open sea. Some of them move in immense swarms, miles in extent. Figures 426–428.

They much resemble the young of ordinary Sea Snails. They form the principle food of the Right Whale. One kind, the Clio, Figure 428, is said to have upon the head three hundred and sixty thousand suckers!

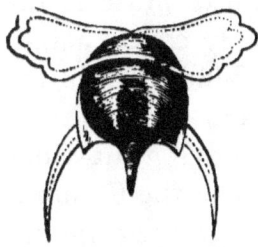  

Fig. 426. — Hylea. Atlantic.   Fig. 427. — Limacina. South Polar Seas.   Fig. 428. — Clio. Arctic.

LAMELLIBRANCHIA.

These have no distinct head, breathe by plate-like gills, usually two pairs, and have a bivalve shell.

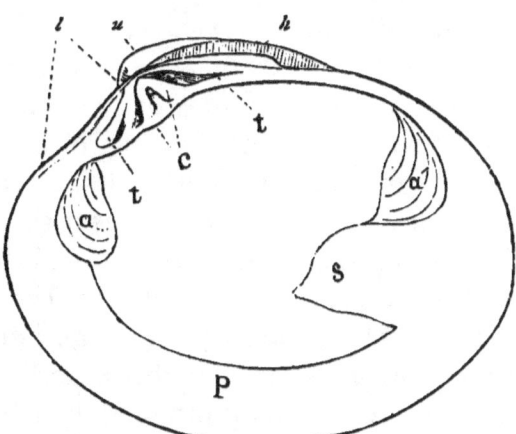

Fig. 429. — One valve of a Bivalve Shell, showing the names of the parts.

a, anterior retractor muscle; *d*, posterior retractor; *t*, lateral teeth; *c*, cardinal tooth; *l*, lunale; *u*, umbo; *h*, hinge ligament; *s*. retractor of siphons; *p*, pallial impression.

These Bivalves have a shell composed of two pieces, or valves, joined together on one side by a hinge, and held tightly together by one or two strong muscles which pass from one valve to the other on the inside. When the animal relaxes these muscles the shell is forced open by an elastic body called a ligament, situated at the hinge. Some kinds live in the sea, others in brooks, rivers, ponds, and lakes. Some idea of them all may be gained by studying the Common Mussel, Figure 437, of the brooks, or the Common Clam, Figure 450, of the seacoast. Place the clam in a large basin of sea water, and soon it will begin to put out a dark-colored organ as long as the shell. The Clam can stretch it out two or three times the length of the shell. This is supposed by many persons to be the head, but it is not; the mouth is within the shell and at the opposite end. At the end of the dark organ are two holes,—one larger than the other,—these being the openings of two tubes which are inclosed in the dark-colored sheath; and around each opening there is a row of fringes or tentacles. A current of water is all the time flowing into the larger opening, and another current flowing out of the smaller opening. The first carries in pure water to supply air to the gills, and minute plants and animals to supply the mouth and stomach with food, and the outgoing current bears away the impure water together with the waste particles which the animal throws off. The currents are caused by a vast number of hairlike fringes which cover the gills within the Mollusk, and which are constantly in motion. The position and appearance of the siphonal tubes in fresh-water Mussels are seen in Figure 435.

Though mainly small, or of ordinary size, a few Bivalves are very large. At St. Sulpice, Paris, the valves of a Tridacna, two feet across, are used as vessels for the holy water. The Tridacna lives on coral reefs in the Pacific and Indian Oceans.

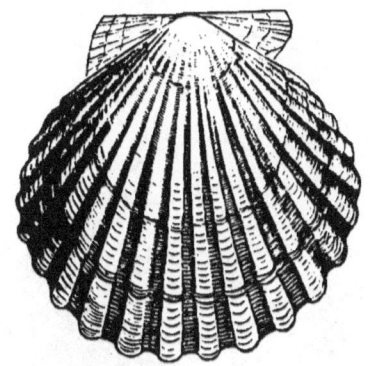 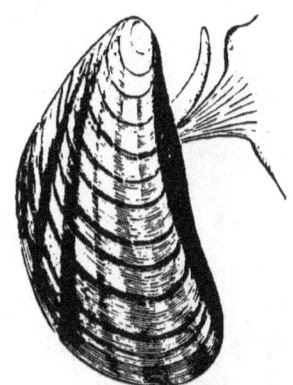

Fig. 430. — Pecten. From Cape Ann southward. Fig. 431. — Mytilus. Both shores of the Atlantic.

 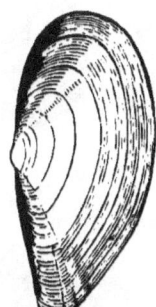 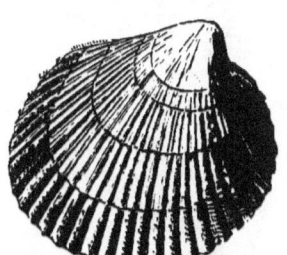

Fig. 432. — Avicula. Reduced. Mediterranean. Fig. 433. — Leda. New England. Fig. 434. — Cardicum. Reduced. New England.

Oysters, Pectens, Mussels, Pearl Oysters, etc.

Oysters are more highly prized for food than any other mollusk. They occur in the greatest quantities

MOLLUSKS: LAMELLIBRANCHIA.

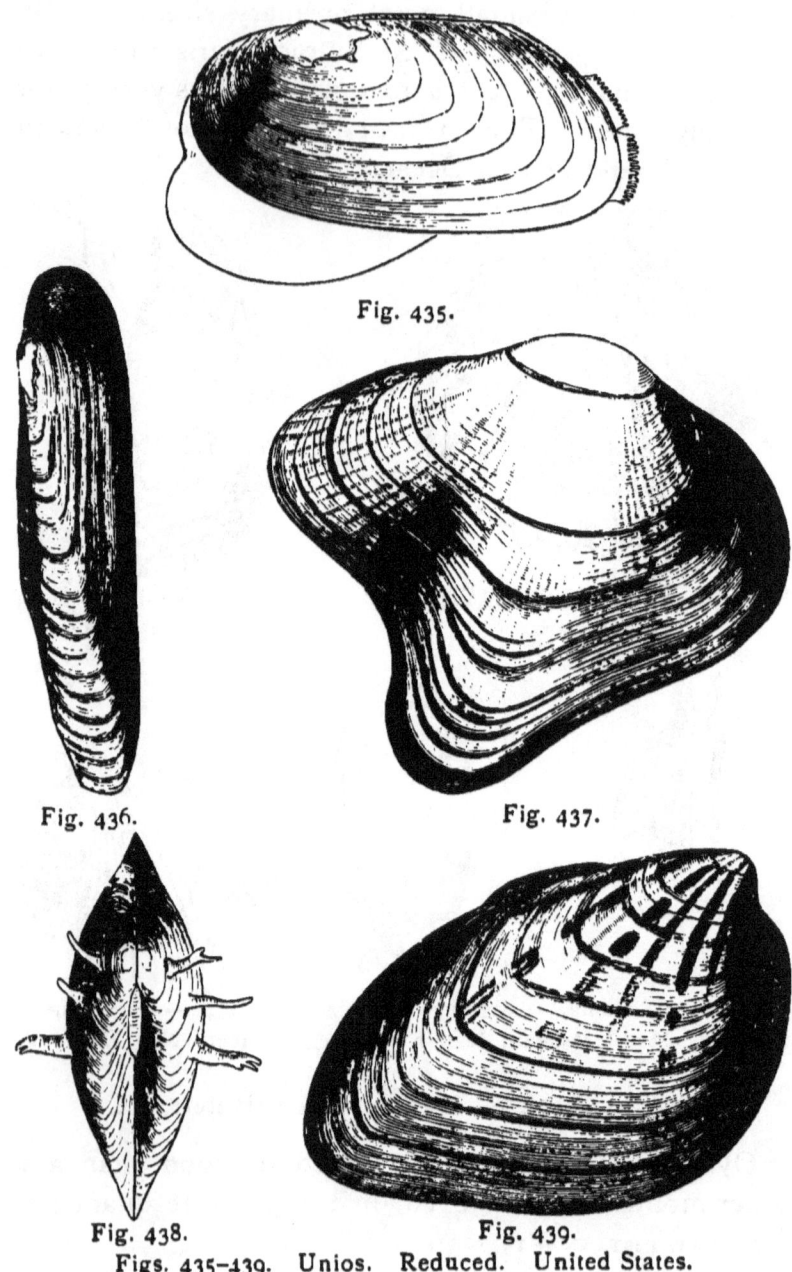

Fig. 435.
Fig. 436.
Fig. 437.
Fig. 438.
Fig. 439.
Figs. 435–439. Unios. Reduced. United States.

on the coast of the Middle States, especially in Delaware and Chesapeake Bays.

Pectens, or Scallops, Figure 430, are also prized for food. Their beautiful shells are known to almost every one, for they are much used in making card-holders, pin-cushions, and other little articles. The Pecten swims rapidly by opening and shutting its valves.

Sea Mussels, Figure 431, inhabit mud-banks which are uncovered at low water. They multiply rapidly, and grow to their full size in one year. By means of a collection of horny threads, called a *byssus*, they attach themselves to rocks, or to the ground.

Pearl Oysters, or Aviculas, Figure 432, have shells yielding the beautiful material called *mother-of-pearl*, extensively used for making and ornamenting a great number of useful and beautiful articles. They also yield the oriental pearls.

Unios, or River Mussels.

These Mollusks abound in brooks, rivers, ponds, and fresh-water lakes. They are sometimes called Naiades, and there are very many kinds. It would take several books larger than this one to describe all the kinds found in the United States. A few of the forms of Unios are shown in Figures 435–439. Sometimes beautiful and valuable pearls are found in them.

Razor Shells, Clams, etc.

The Razor Shells are very long and smooth. They burrow in the sand, and are good for food. The Common Clam burrows in sand and mud, and is extensively used for food, and for bait for cod.

Fig. 440. — Astarte. New England.

Fig. 441 — Sphærium. Northern States.

Fig. 442. — Thyasira. New England.

Fig. 443. — Sphærium. Northern States.

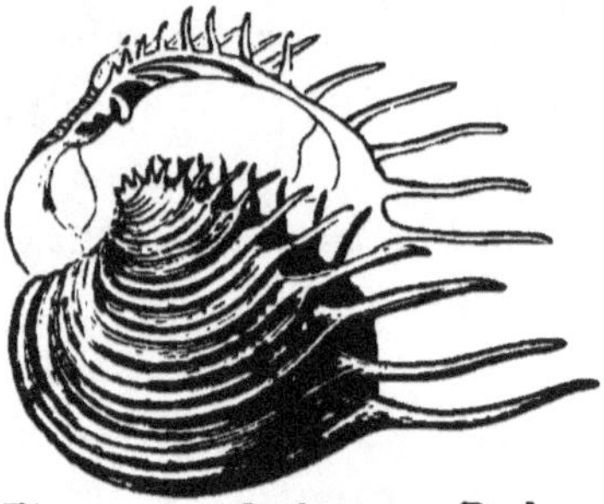

Fig. 444. — Cytherea. Reduced. West Indies.

Fig. 445. — Mactra. Great Britain.

Fig. 446. — Tellina. Great Britain.

Fig. 447. — Tellina. United States.

Fig. 448. — Tellina. United States.

Fig. 449. — Razor Shell, or Solen. Much reduced. Both shores of the Atlantic.

Pholades and Shipworms.

Pholades have the shell very hard and rough, like a rasp, and they burrow into all sorts of substances, even into stone. Shipworms are long Mollusks, looking like

Worms. The common kinds are about a foot long, but one kind is three feet in length. They bore into the timber of ships and wharves and are very destructive.

Fig. 450. — Common Clam. Reduced. Coast of New England.

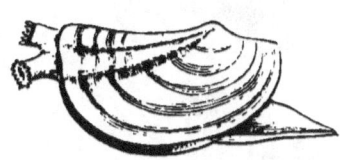

Fig. 451. — Pandora.

Fig. 452. — Gastrochæna. Galway.

Fig. 453. — Watering-pot Shell. Much reduced.

Fig. 454. — Pholas. Reduced. India.

Fig. 455. — Pholas. New England and eastward.

VERMES, OR WORMS.

Worms are long animals, which are made up of many similar rings. The nerves are distributed in knots or clusters throughout the whole length of the animal, and hence worms are not always killed when cut in pieces; and in some cases the several pieces become distinct worms. The kinds of Worm are very numerous, but they are most abundant in the sea and in fresh waters. Many of the animals which look like Worms are Butterflies or Moths, in the larva state, and are caterpillars instead of Worms. Such are the "Tobacco Worm" and "Tomato Worm." One of the most common kinds of Sea-worm is the Serpula, which lives in tubes that are found incrusting stones and other bodies. The breathing organs are in tufts near the head, and there is a little round body, shown in the cut, which serves to shut the animal in when it withdraws itself into the tube.

Fig. 456.—Serpula.

The Earthworm, common in rich soils, is well known to all boys, and is used as good bait for trout and other fishes. It feeds upon the tender roots of plants and also on their leaves. It may swallow earth for the sake of the minute animals and plants in it. In spite

Fig. 457.—Earthworm.

of the absence of legs, jaws or other hard parts, which can be employed in digging, the worm is able to drive long burrows even in firm soil. As every boy knows, it can fasten itself firmly into its burrow by shortening and so thickening that part of the body which is in the burrow. In tunneling, the same means is used to prevent the worm from slipping backward while the forward end of the body is forced into the unpierced ground. The worm may swallow the earth if it is too hard to be pierced in this way. The earth thus swallowed or taken in with the food is discharged on the surface of the ground. These "worm-castings" are very abundant after rains or in the morning after a damp night, as the worms are very active at night and in damp weather. Worms play an important part in causing changes in the soil. Their burrows allow the air and water to easily penetrate the earth and the worms carry very large amounts of soil to the surface from the deeper parts of the ground. The Worm is by no means the slow and torpid creature which most of us fancy it to be. At night they move about actively on the surface of the ground, after their chief enemies, the birds, are asleep. They can also climb, as their presence in eaves, troughs and water tubs shows. The senses of the Worm are not highly developed. Touch and taste are perhaps the best. They show a sort of intelligence in the arrangement of leaves to plug their burrows. The leaf is always drawn into the burrow by the narrow end and the worm feels the leaf all over in order to seize it by the right end.

In tropical countries Earthworms are found much larger than ours. Some of these are as much as six feet long and thick in proportion. Many small Worms, similar to the Earthworm, are found in fresh water.

BRACHIOPODS.

Some Worms so differ in form from the ordinary Worm that they are hardly recognizable as Worms. Such are the Brachiopods and Polyzoa.

These animals have the two valves of unequal size,

and in one of them there is a hole through which passes a fleshy stalk, by which the shell is attached to the rocks. The word Brachiopod means *arm-footed*, and is given to these animals on account of the long, fringed arms growing from the sides of the mouth,

Fig. 458.— Terebratula — a Brachiopod.   Fig. 459.— Side view of Fig. 458.   Fig. 460. — Brachiopod.

Fig. 461.—Lingula — a Brachiopod.   Reduced.   Philippines.

and by means of which they make currents in the water and thus secure their food.

The Lingula has a very long fleshy stem and so shows its relation to the Worms more plainly than the rest. The shell of this animal is found as a fossil in the most ancient rocks which contain traces of animal life.

POLYZOA.

These are very small, or minute, animals related to the Worms, growing in clusters upon shells, rocks, and other objects, both in the sea and in fresh waters, and which look very much like Polyps. They are often called Bryozoa.

PARASITIC WORMS.

There are a great number of Worms which live as parasites in the intestines, or in the flesh of other animals. Such are the Tapeworms and the Threadworms. The Trichina is the most dangerous of these parasites. It lives in the flesh of the hog, and if eaten by man the worms develop in the intestine and afterwards burrow out into the muscles. Their migration causes great suffering and often death.

The Gordius, or Hair Worm, belongs to the same great group of Worms as the Trichina. It lives, when young, in the intestine of the Cricket and Grasshopper but becomes free when grown up. It lays its eggs in the water, where it is often found and sometimes thought to be an animated hair.

ECHINODERMS, OR STARFISHES.

The word Echinoderm means *hedgehog-skin*, and is given to these animals because many of them have the outside covered with spines, thus reminding us of the Hedgehog of the fields, described on page 60. In the Echinoderms the parts are arranged according to a reigning number, generally *five;* that is, the parts of each kind are five, or a multiple of five.

SEA CUCUMBERS, OR HOLOTHURIANS.

The Holothurians, or Sea Cucumbers, have no spines, but are covered with a tough skin capable of great expansion and contraction, and containing particles of limestone. There are many kinds, varying from an

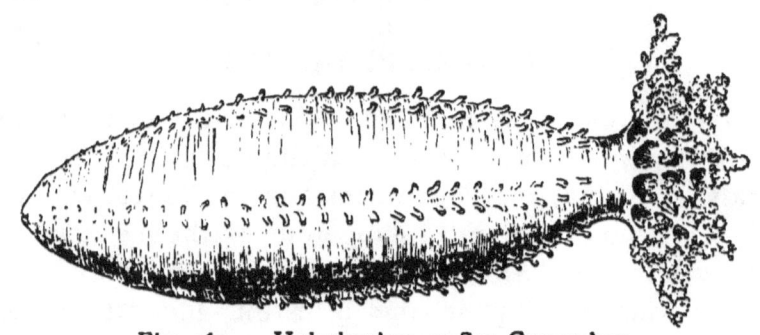

Fig. 462. — Holothurian, or Sea Cucumber.

inch to several feet in length. They live in the sea and are exceedingly interesting, and very beautiful when the long and delicate fringes around the mouth are expanded. When taken from the water they shrink and lose their beauty of form and color. They must be seen in the ocean, or in the aquarium, in order to get a good idea of them. Figure 462 shows one very common at Grand Manan, Eastport, and other places on the North Atlantic. The Chinese use their dried and smoked skins, called *trepang*, in making soup.

SEA URCHINS, OR ECHINOIDEA.

True Sea Urchins are hemispherical, or flattened, and have a hard shell composed of plates which are regular

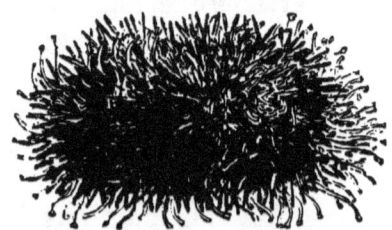

Fig. 463. — Sea Urchin.

in form and firmly bound together. Upon these plates are tubercles, and on these tubercles hard spines. In certain plates there are rows of holes through which pass fleshy organs called suckers, or ambulacral feet, with the end slightly expanded. By means of these suckers, which can be extended much beyond the spines, these animals can cling firmly to other bodies, and thus move about over the rocks, even up and down their smooth sides, as well as on level surfaces. So much can these suckers be extended that a Sea Urchin has been seen to put them forth from the top, and, bending them downwards, cling to the bottom of the basin in which the animal was lying. Figure 463 shows a common kind of Sea Urchin as it appears when alive. When the animal dies, the skin, which covers the shell and holds the spines in their places,

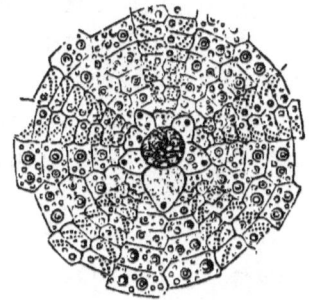

Fig. 464. — Top view of Sea Urchin. Spines removed.

dries up, and the spines fall off, and then the shell, with all its beautiful structure and markings, is plainly seen. In the one represented in Figure 464 we find ten double rows of plates which run along the curved surface from the bottom to the top of the shell. In five of these double rows the plates are large, without holes, and are covered with large tubercles. Alternating with the double rows of large plates are five double rows of smaller ones, bearing few and small tubercles, and each plate is perforated with the holes for the suckers. The plates which bear the holes are called the *ambulacral* plates, — from a Latin word which means a *walk*, or *alley;* and the large plates without holes are called the *interambulacral* plates. At the termination of each of the five belts or zones of ambulacral plates there is a little triangular plate with a minute opening which marks the place of the eye. Alternating with these *ocular* plates, so-called, are five larger plates, each being perforated with a larger hole through which the eggs are laid. One of these plates is much larger than the others, and is filled with very minute holes, and is called by naturalists the *madreporic* body. It serves as a filter or strainer to the water which passes through it into the water tubes of the animal and which is used in extending the sucker feet. The mouth, at the under side, is armed with five strong pointed and polished teeth, which form the outer part of a remarkable dental apparatus, which is called Aristotle's Lantern. In a Sea Urchin of ordinary size there are five or six hundred plates, all fitting together in the most perfect manner, and bearing more than four thousand spines; and the suckers number nearly two thousand.

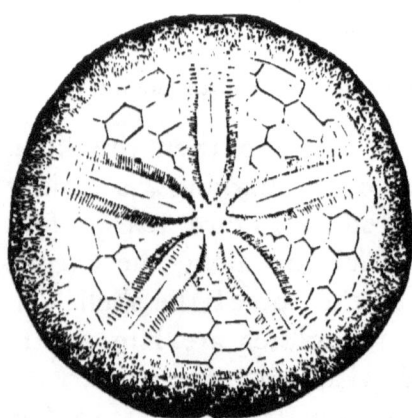

Fig. 465. — Echinarachnius. Northeast coast of North America.

Besides the spines and the suckers, there are scattered over the body and around the mouth of Sea Urchins a great

SEA URCHINS.

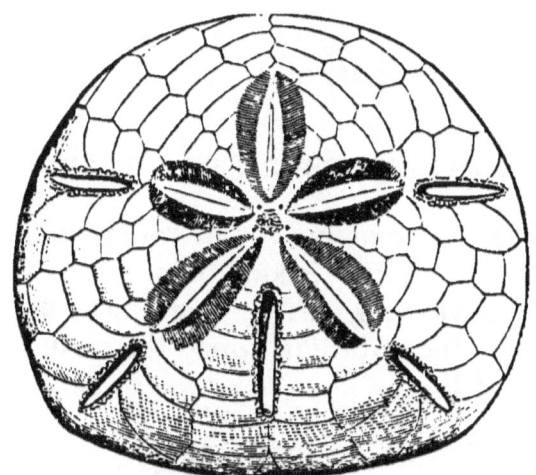

Fig. 466. — Mellita. Southeast coast of United States.

number of curious little organs called *pedicillariæ*. They look like a stem ending in a knob, but the knob is composed of three pieces or blades, which open and shut tightly, thus forming a sort of pincers. These organs are of use in keeping the shell clean.

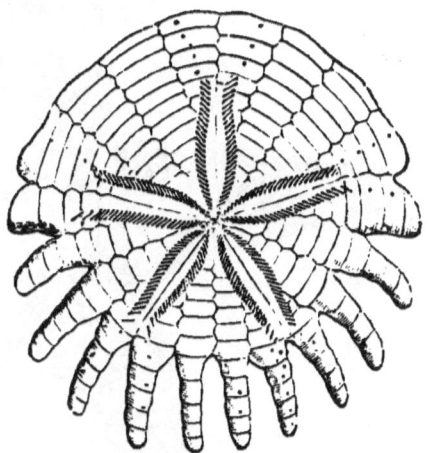

Fig. 467. — Rotula. Coast of Africa.

The number of kinds of Sea Urchins is quite large, and they vary in size from an inch to three or four inches in diameter, and have spines from a quarter of an inch long to three or four inches in length. Some of them are capable of making holes in hard substances, even in limestone and granite.

Other kinds, like Figures, 465, 466 and 467, burrow in the sand. These are much flattened.

STARFISHES, OR SEA STARS.

Starfishes are common on all rocky coasts. They are readily found by looking under the seaweed in pools that have been left by the tide. They are so named from their starlike form, the disk or central

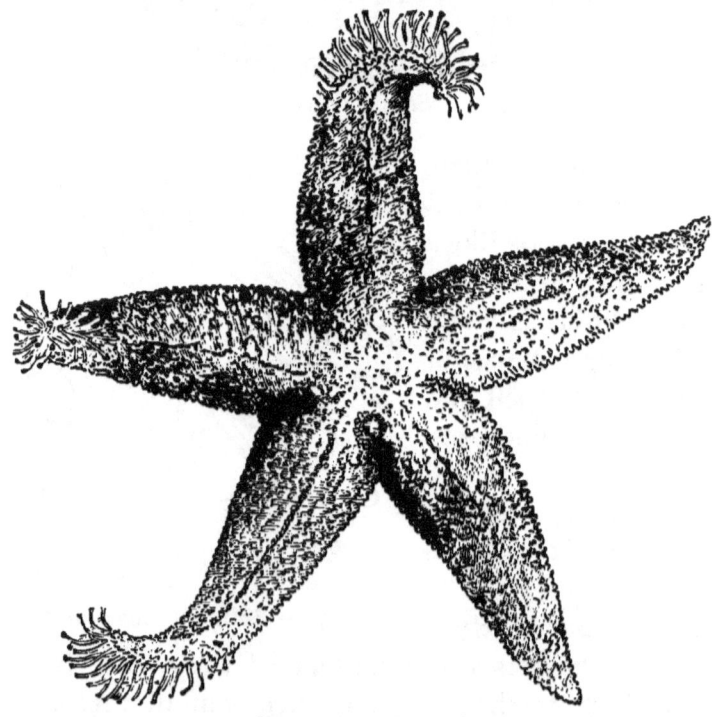

Fig. 468. — Starfish.

portion gradually merging into the rays. Beneath each ray there is a large number of locomotive suckers, like those of the Sea Urchins already described.

These tubes are seen in Figure 468, which shows the upper part of the Starfish, three of the rays being slightly turned backward. The mouth is on the under side in the center, and there is an eye, or eye-spot, at the end of each ray. By means of the ambulacral tubes Starfishes move slowly but surely over the rocks and all kinds of surfaces, and they can cling to the rocks so firmly that they are often removed with difficulty, and will sometimes even allow their ambulacra to be pulled off rather than let go their hold. Their covering is not solid as in the Sea Urchins, but is composed of movable plates, so that these animals are able to bend themselves in every direction, and thus work their way into holes and fissures in rocks where we should hardly expect to find them. Starfishes feed upon mollusks and other marine animals, and when they feed they turn the stomach out of the mouth and over the food to be devoured. A curious spot is seen on the back near the junction of two of the arms. This is the madreporic body described in speaking of the Sea Urchins. It is a sort of minute sieve, and forms an entrance to a series of internal water-tubes, some of them connecting with the locomotive suckers and supplying them with water. Water is also admitted into the body through minute pores which cover the whole surface of the animal. Starfishes often lose one or more of their arms, or rays, by being dashed against the rocks by the waves, or the arm is bitten off by a fish. In all such cases a new one sprouts out in the place of the old one, and specimens may be found showing such new rays in all stages, from those that are just sprouting to those that have nearly reached their full growth.

SERPENT STARS, OR OPHIURANS.

The Serpent Stars, or Ophiurans, are so called from the resemblance of their long slender rays to a snake's

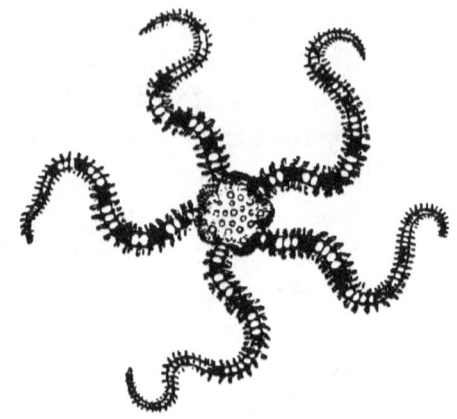

Fig. 469.— Serpent Star, or Ophiuran.

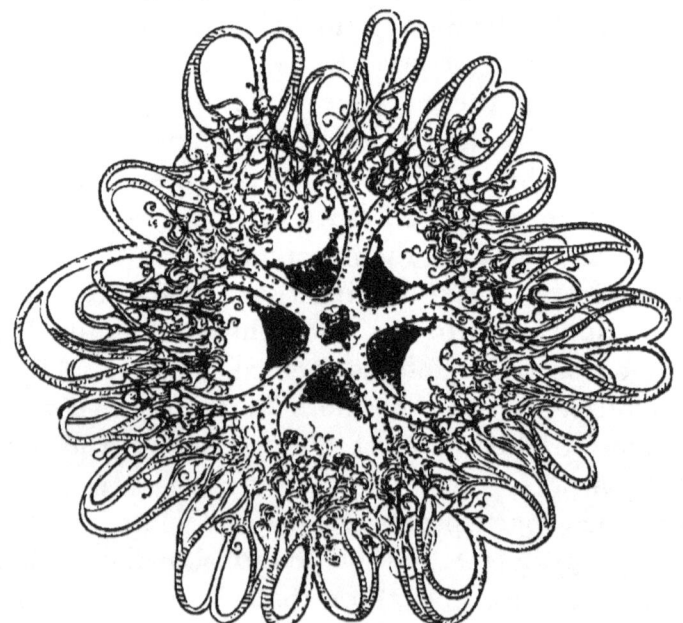

Fig. 470. — Basket Fish, or Astrophyton.

tail. They are found on nearly all coasts, and are at once distinguished by a small disk or central portion from which the rays start off very abruptly, instead of the gradual passage of the central part into the arms, as seen in the true Starfishes. They move about by bending their arms, whose hold is aided by their spine. Nearly all have the arms simple, Figure 469; but some have the arms much branched, Figure 470.

CRINOIDS.

The word Crinoid means *lily-like in form*, and is given to a large number of Echinoderms on account of their lily-like or plant-like appearance. Only a small

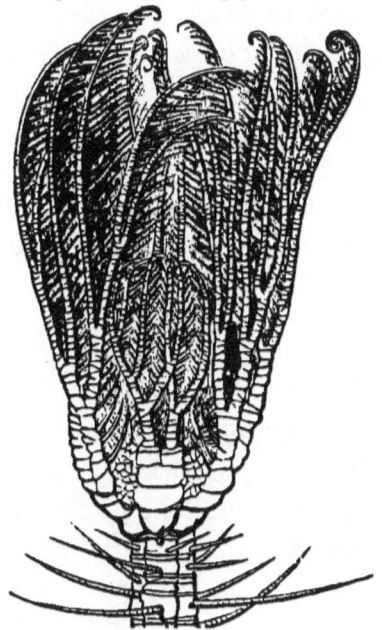

Fig. 471.—Living Crinoid—Pentacrinus caput-medusæ. West Indies.

number of these animals are now living. Of the few living ones, some kinds have a stem, in the adult state.

NAT. HIST. AN.—16

One of these is the *Pentacrinus caput-medusæ*, of the West Indies, Figure 471. Some of the living Crinoids are free-swimming when adult, and much resemble the Ophiurans.

In the rocks, in various parts of the United States and in other countries, fossils of the stemmed kinds are exceedingly abundant, showing us that these animals lived in profusion in the old ocean which ages ago covered a large part of our country. And the fossils are so various in form, and so beautiful in pattern and marking, that no words can fitly describe them. The workman in the quarry stops to admire them, and the learned naturalist is fascinated by their beauty, and never grows weary of studying them. They are the gems of the geological collection.

Of the living free Crinoids, — that is, those without a stem, — one of the best known is called the Antedon, or Feather Star. When young this, too, has a stem, and looks not very unlike the Medusa's Head, Figure 471; but as it grows older it drops from the stem, and lives a free life.

CŒLENTERATES, OR LASSO-THOWERS.

These radiated animals are cylindrical in form. They have no digestive canal separate from the body wall. All of them have nettle cells, as described below.

The main kinds are the Jellyfishes including Ctenophora and Hydroids, and Polyps.

JELLYFISHES, OR ACALEPHS.

Of all animals of the sea, perhaps none are more wonderful than these. Their jellylike bodies, curious forms and structure, their beautiful colors of claret, rose and pink, their varied and almost magical movements, as varied and graceful as those of the birds and insects of the air, their phosphorescence by night, causing them to be called the "lamps of the sea,"

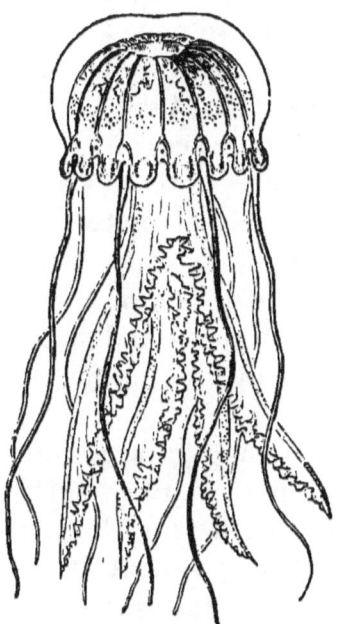

Fig. 472. — Jellyfish.

and their curious changes in passing from the young to the adult state, have interested all intelligent visitors to the seaside, and have caused these animals to be carefully studied by some of the most eminent naturalists of Europe and America. The word Acaleph means *nettle*, and is given to these animals because some of them cause a stinging sensation when they touch our flesh; hence they are often called Sea Nettles. They are also as often called Medusæ. Their common name, Jellyfish, was given on account of their jellylike appearance and substance.

If we examine the structures of Acalephs, we find a cavity, which is the stomach, hollowed out of the mass of the body, and this cavity has an opening which serves as a mouth; the edges of this opening are turned outward and prolonged into delicate fringes. And there are tubes which radiate from the center of the body and unite with a tube at the circumference.

The kinds of Jellyfishes are numerous, and they vary in size from those scarcely visible to those which are one or two yards in diameter, and with tentacles thirty or forty feet long; and Mrs. Agassiz, in her beautiful book, "Seaside Studies in Natural History," mentions one which measured about seven feet in diameter, and had tentacles more than a hundred feet in length!

Jellyfishes are a hungry race, and feed upon their own kind and other marine animals, which they secure by means of their tentacles and lassos. On the tentacles of Jellyfishes, and of Polyps too, there are numerous lasso-cells, — too small to be seen without the microscope, — each containing a long, spirally-coiled thread or lasso, which can be instantly darted forth and thrust into the little animal which is desired for food.

CTENOPHORA.

The Ctenophora are more or less spherical, or egg-shaped, with eight rows of locomotive fringes dividing the surface of the body, as the ribs divide the surface of a melon. The Pleurobrachia is one of the most common kinds on the northeast coast of the United States, and in its movements and curious appendages is one of the most wonderful of all the Jellyfishes. It is transparent, and besides the eight rows of fringes mentioned above, it has two most extraordinary tentacles one on either side of the body; and no form of expansion or contraction, or curve or spiral, can be conceived which these tentacles may not assume.

Bolina and Idyia are other Ctenophora common on

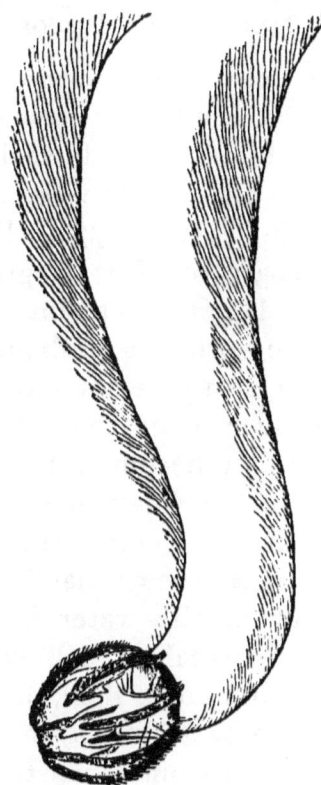

Fig. 473. — Pleurobrachia.

the northeast coast of the United States. The Rose-colored Idyia is three or four inches long, and shaped somewhat like a melon with one end cut off. The mouth occupies the whole of the cut-off end, and the stomach occupies a large part of the interior of the animal. In summer it sometimes appears in such swarms as to tinge large patches of the sea with a delicate rosy hue. It is very voracious, and feeds mainly on other jellyfishes, sometimes capturing those nearly as large as itself.

TRUE MEDUSÆ, OR DISCOPHORA.

These have the body in the form of a hemispheric disk, more or less flattened. Of these disk-shaped Medusæ none are more beautiful in their appearance or interesting in their history than the Aurelia, or Sunfish, represented in Figure 477. This Jellyfish is common on the coast of New England, is about a foot across in the larger specimens, and lives but a single year. In the spring it is about a quarter of an inch in diameter, and on pleasant days moves in large swarms near the surface of the water. About the middle of summer they become full grown. Towards the close of summer they lay their eggs, and in the autumn they perish. At length the eggs hatch, and the little *planulæ*, as the newly hatched Jellyfishes are called, swim about in the water by means of tiny appendages which naturalists call *cilia*. Soon each becomes attached to a rock, shell, or seaweed, and is then called *scyphostoma*, Figure 475. Then the body begins to divide by horizontal constrictions, and soon appears as in Figures 474 and 476, and is then called *strobila*. At length the segments become more and more separated, and the uppermost one drops off, then the next one, then the next, and so on till each in turn has separated from the one below itself. Each disk, as it separates, turns over and floats away, and is known as *ephyra*. Soon each ephyra assumes the form of a

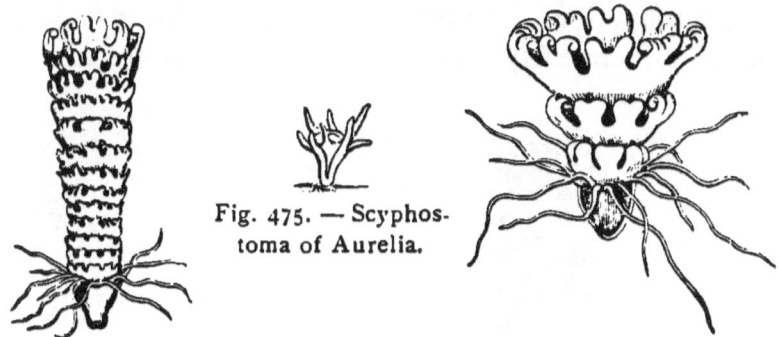

Fig. 475. — Scyphostoma of Aurelia.

Fig. 474. — Strobila of Aurelia. Magnified.

Fig. 476. — Strobila of Aurelia. Much magnified.

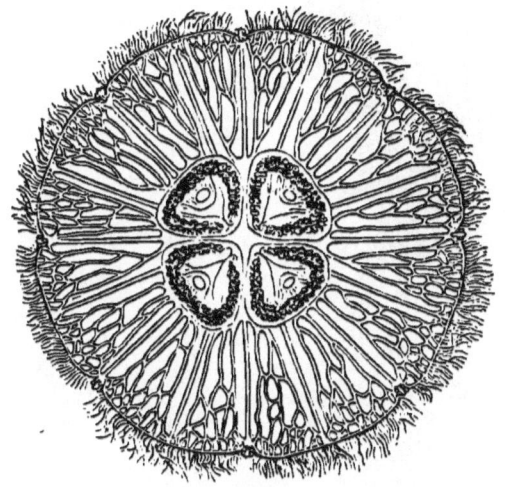

Fig. 477. — Sunfish, or Aurelia.

perfect Jellyfish, as shown in Figure 477. Thus one scyphostoma which comes from a single egg becomes a strobila, and this strobila divides into numerous parts, each of which becomes a Jellyfish.

248   CŒLENTERATES: ACALEPHS.

HYDROIDS.

The Hydroids are Jellyfishes which are almost more wonderful in their mode of development than those already described. Occurring, as they do in many cases, in their early stages of existence, as mere dis-

Fig. 478. — Coryne. Cluster of Hydræ growing on seaweed.

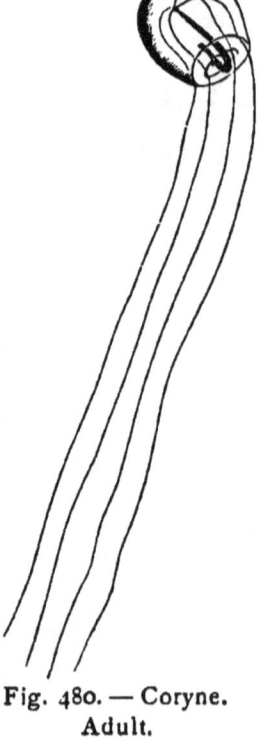

Fig. 480. — Coryne. Adult.

Fig. 479. — Single individual of Fig. 478, enlarged, showing *a* and *b* just ready to drop off and become free Medusæ, like Fig. 480; *c*, a younger bud.

colored patches on seaweeds, stones, or shells, or appearing like little tufts of moss, or miniature shrubs, the untrained eye might well mistake the fact that they are animals. But naturalists have shown that these plantlike forms produce Medusæ buds, which expand into genuine Medusæ, or Jellyfishes. Figure 478 shows a little cluster of Hydroids attached to seaweed, and Figure 479 shows a single individual of the same very much magnified, with two of the buds much enlarged, and a third quite prominent. Soon each bud becomes detached, and floats away as a free Jellyfish,

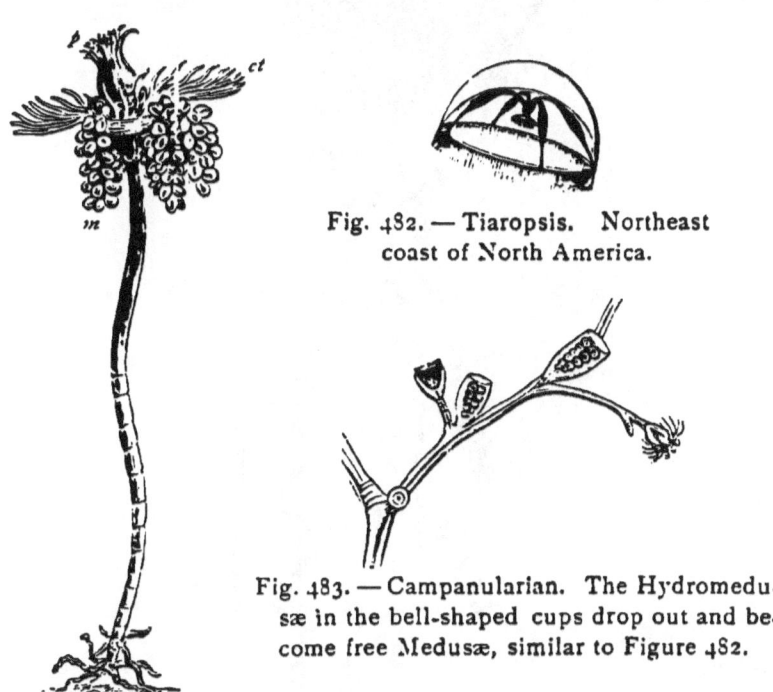

Fig. 482. — Tiaropsis. Northeast coast of North America.

Fig. 483. — Campanularian. The Hydromedusæ in the bell-shaped cups drop out and become free Medusæ, similar to Figure 482.

Fig. 481. — Tubularia. Massachusetts Bay.

m, medusæ; *ct*, coronal tentacle; *p*, proboscis.

like Figure 480, and is then known as Coryne, or, as it was formerly called, Sarsia, so named from Sars, a

Norwegian naturalist, who was one of the first investigators of these curious kinds of Jellyfishes.

Nothing can excel the delicacy of Coryne. Soft as the softest jelly, almost as transparent as the dewdrop,

Fig. 484. — Portuguese Man-of-war.

yet it performs varied and rapid movements, contracts and expands its tentacles, catches and devours other Medusæ, and other marine animals, and to all appearances delights in life as much as higher animals do.

They are abundant in the spring. In the middle of summer they lay their eggs and perish. But the eggs do not hatch Medusæ like the parent, but each hatches a little hydroid which is first free, then afterwards becomes attached to a shell, seaweed, or stone, and from this little hydroid others branch till a little community of hydroids had grown up, as in Figure 478. From these hydroids bud again the Coryne, Figure 479.

In some kinds, as Tubularia, Figure 481, the hydroid has a wreath of coronal tentacles, as they are called, a projecting part called a proboscis, and the medusæ grow in clusters from just above the coronal tentacles.

In those called Sertularians and Campanularians, Figure 483, the hydroid has a stem which is covered by a horny sheath, forming a cup around the head. In a fertile cup there are a dozen or more hydromedusæ, which at length drop out and become free medusæ similar to Tiaropsis, Figure 482.

In those called Siphonophora, the hydroids exist as free moving communities, each community being made up of individuals of different kinds, yet all so combined as to give the appearance of one animal. The Portuguese Man-of-war, of the Gulf of Mexico, is one of the most remarkable and best known of this sort. It consists of a pear-shaped and elegantly crested air sac, floating lightly upon the water, and giving off from its under surface numerous long and varied appendages. These are the different members of the community, and fill different offices; some of them eat for the whole, others produce medusa buds, and others are the locomotive or swimming members, and have tentacles that stretch out behind the floating community to the length of twenty or thirty feet.

CŒLENTERATES: POLYPS.

Fig. 485. — Acalephian Coral.

It was discovered by Professor Agassiz, that there are some kinds of Acalephs which produce coral similar to that formed by Coral Polyps, described in the following pages, but unlike the latter in having, in the cells, a horizontal floor extending from wall to wall.

SEA ANEMONES AND CORAL ANIMALS, OR POLYPS.

These are marine Cœlenterates which have a sacklike or tubular body, with a circular top, in the center of which is an opening called the mouth, and around the

Fig. 486. — Polyp — Sea Anemone.

mouth are one or more rows of hollow feelers, or tentacles. The mouth opens directly into an inner sack, which is the stomach, and this stomach opens at the bottom into the main body. The main body is divided by partitions, which run from the bottom to the top, and from the outer wall to the stomach. Through the opening at the bottom of the stomach there is free communication with all the chambers formed by the partitions, and these chambers connect with the tentacles; so that the food, after being digested, passes into the main body, and thence into the tentacles, thus nourish-

Fig. 487. — Cluster of Coral Polyps in various stages of expansion.

ing every part. The food of polyps consists of small marine animals of various kinds, which are secured by means of the tentacles and the curious and wonderful lassos situated on the tentacles. The word Polyp means *many-footed*, and is given to these animals on account of their numerous tentacles; but it must not be supposed that the latter are feet in any true sense. Most kinds of Polyps are attached to the rocks, shells, or other bodies beneath the waves. Some live singly, others in communities whose numbers are often far more numerous than the leaves upon the trees.

Polyps increase by means of eggs, by budding in a manner much like that of trees and shrubs, and by division of one animal into two or more, so that the largest communities arise from a single animal. The eggs are formed on the vertical partitions, and pass out, through the mouth, into the water. When first hatched, the young do not look like the parent, but are little oval bodies which move freely about by means of the fringe-like appendages, called *vibratile cilia*, with which they are covered. At length each becomes attached to a rock, shell, or seaweed, and soon assumes the form of the parent. If it is a kind which buds, there soon grow from its sides or base others exactly like itself, and from these, in turn, bud other polyps of the same kind. Thus the community goes on growing till it has reached its limits of increase. If it is a kind which increases by division, it widens as it grows upward, the walls in two opposite places begin to approach each other, and soon the polyp is divided into two, with two mouths and two circular disks surrounded by tentacles, instead of one as before the division. The polyps thus formed divide in the same way, and this process is continued till from a single polyp there is formed a large and beautiful cluster.

Polyps readily reproduce lost parts, and even if cut in pieces, each fragment will, in some cases, become a perfect animal. Polyps vary in size from extreme minuteness to those that are more than a foot across. Some, like the Sea Anemones, Figure 486, are wholly soft; others secrete a more or less solid framework, which is called *coral;* and those which secrete coral are called Coral Polyps, or Coral Animals. Some persons suppose that coral is something that is built by an insect, as the bee builds comb, or the wasp its nest, and the industry of this supposed insect is often spoken of. But it is not proper to give the name Insect to the Coral Polyps, for they are in no way related to Insects, either in appearance, structure, or habits. Coral is not something which is built, but something which grows. It is the skeleton, or many united skeletons, of Polyps, and these animals exhibit no industry in forming it, any more than do other animals in forming their own bones. Coral is not a house in which the animal lives; on the contrary, the coral is wholly inside of the animals, and it is only when the Polyps die, wither, and disappear that we see the solid coral itself. Polyps grow in various and most wonderful and beautiful forms, imitating almost all kinds of vegetation, as lichens, fungi, mosses, ferns, grasses, herbs, shrubs and trees. A hundred years ago, or more, they were thought to be plants, and even the great naturalist, Linnæus, regarded them as plant-animals, that is, partaking of the character of both plants and animals; even now they are often called Zoöphytes, a word which means *animal-plants*, although they are in no way related to plants. The colors of these wonderful animals of the sea are as beautiful and almost as varied as their forms; and some of the Polyp communities equal, in splendor of colors, the most beautiful flower gardens of the land.

Sea-pens, Gorgonias, etc., or Alcyonaria.

These are Polyps which have eight long fringed or lobed tentacles, around a narrow disk, Figures 488–

CŒLENTERATES: POLYPS.

Fig. 488. — Renilla.

Fig. 489. — Single Polyp of Renilla. Enlarged.

Fig. 491. — Sea Fan. Portion of large frond.

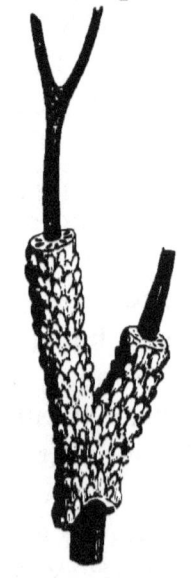

Fig. 490.—Single Polyp of Red Coral. Enlarged.

Fig. 492.—Verrucella.

Fig. 493.— Red Coral.   Fig. 494. — Primnoa.   Fig. 495.—Organ-pipe.

490, and which form compound clusters or communities by budding. The Sea-pens and Renillas are Polyps which are arranged on a more or less expanded disk, which is connected with a sort of stem or peduncle, by means of which the community may move about or fix itself in the sand or mud. The Sea-pens are so called from their resemblance to a quill. The Renilla, Figure 488, found on the coast of the Southern States and of South America, looks like a broad leaf attached to its leaf-stalk; and when the purple disk is covered with the expanded Polyps, as seen in the cut, it is a very beautiful object. The form of a separate Polyp is shown in Figure 489.

The Gorgonias abound in tropical seas, but some kinds are also found in temperate regions. The forms are exceedingly various, Figures 490–494, and many of them are very delicate and beautiful, often bearing a very close resemblance to plants; in all, however, the Polyps are short, and secrete a solid central axis of coral. This axis is plainly shown in Figures 492 and 494. One of the most common and striking forms of the Gorgonias is the Sea Fan, which is more or less broad and fan-shaped, the branches in many cases running together so as to form a network. Figure 491. One form of the Gorgonias, the Primnoa, Figure 494, is found even as far north as St. George's Banks and the Bay of Fundy. But the one which has the greatest popular interest is the Red Coral, *Corallium rubrum*. It is obtained mainly in the Mediterranean. The coral fishers go out in boats, and are provided with a large wooden cross, which is loaded with stone in the center and has hempen nets attached to each of its arms. While one man is constantly raising and letting fall this machine upon the coral beds, others row the boat so that the branches broken off are caught up by the nets. From time to time the cross and nets are raised, and the branches of coral which have been entangled in the meshes are secured.

Closely related to the Gorgonias are the Alcyonacea, of which the Organ-pipe Coral, Figure 495, is one of the most interesting examples. It is of a beautiful red color, and gets its name from the fact that the tubes of the coral somewhat resemble organ-pipes.

Sea Anemones, or Actinaria.

These Polyps are wholly soft, only a few secreting from the base a horn-like substance. They are common on nearly all coasts, and vary from a quarter of an inch to a foot or more in diameter, as seen in some of the tropical species. Our species seldom exceed two or three inches in diameter, and most of them are much smaller, although some are six inches high. The Bunodes, Figure 499, is found among the rocks on the coast of Maine. The most common kind on the northeast coast of North America is the Fringed Actinia, or Metridium. Figures 496–498. When fully expanded, it is about four inches high and three inches across the disk, and is a most interesting object.

Madrepores, Porites, Meandrinas, Astræans, etc., or Madreporaria.

These Polyps are simple or compound, often excessively branching, and they form coral in their walls, or outer parts, in their radiating partitions, and often at their base. The forms which the communities assume

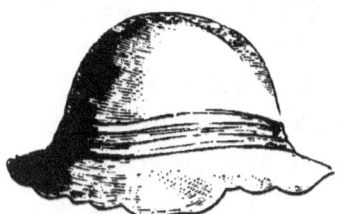

Fig. 496. -- Same as Fig. 498. Closed.

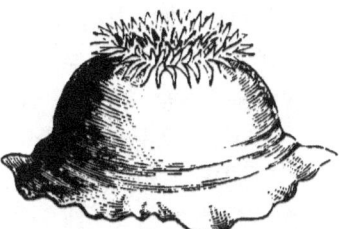

Fig. 497. — Same as Fig. 498. Just opening.

Fig. 498. — Sea Anemone, or Fringed Actinia.

Fig. 499. — Sea Anemone, or Bunodes.

are very beautiful and exceedingly various, and they are among the most beautiful objects in zoölogical cabinets.

The great group of Madrepores contains Polyps which have a definite number of tentacles, twelve or more; those called Porites, Figure 501, have the cells shallow, and are not more than one twelfth of an inch in diameter, the coral in some cases branching, in others massive, and always very solid. Massive specimens of Porites are sometimes fifteen feet in diameter.

In the true Madrepores, Figure 500, the Polyps do not secrete coral at the base; hence the cells of the coral are very deep, and these corals spread and branch into the most beautiful and varied forms, and the Polyp at the end of a branch, Figure 500, is always larger than the others.

In the great group of the Astræans the tentacles occur in multiples of six. Those of this group, called Brain Corals, or Meandrinas, have the surface covered with winding trenches, Figure 505, on each side of which there is a row of tentacles. The form of the Meandrinas is generally that of a hemisphere, and some of these masses are twelve feet in diameter. The true Astræans, or Star Corals, Figure 507, have the cells in the form of concave pits, and the common forms of this coral are hemispherical or dome-shaped masses, some of which are twenty feet in diameter; the Polyps themselves are often an inch in diameter. Most of them, however, are very much smaller. One beautiful little Astræan, Dana's Astrangia, has its home in Long Island Sound, where it occurs in little clusters upon the stones and shells, from just below low-water mark even down to ten fathoms in depth. It thrives well in the aquarium, and eats little mollusks and other small animals with a good relish. In those Coral Polyps called Oculinas, the coral, when young, spreads so as to form a broad base; later, beautiful tufts and treelike branches arise from this base. A portion of one of these Oculinas is shown in Figure 509.

In the great group of Fungus Corals, the coral is broad and flat, looking like a toadstool without a stem, as in Figure 510. Polyps of this kind have

MADREPORARIA.

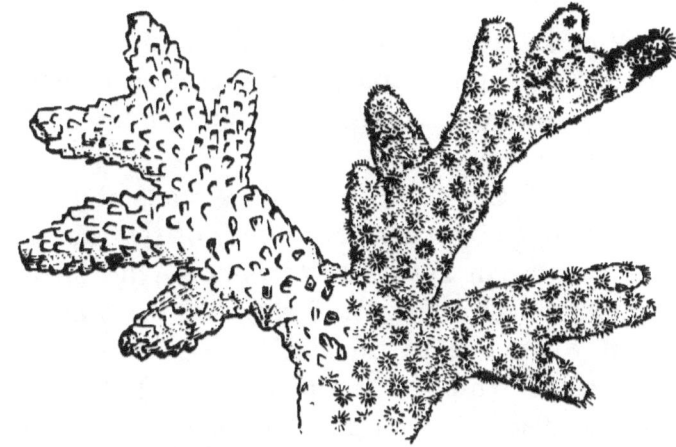

Fig. 500. — Madrepore. Right-hand branches alive.

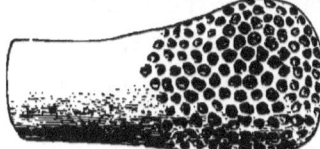

Fig. 501. — Porites.

Fig. 503. — Cœnopsammia.

Fig. 502. — Astroides. Coral Polyps in various stages of expansion.

Fig. 504. — Dry Coral. Same as Fig. 503.

short lobe-like tentacles in multiples of six. Each specimen, like Figure 510, is the secretion of a single Polyp, and similar specimens are sometimes a foot or more in diameter.

But some of the most interesting facts about Coral Polyps remain to be told. Hundreds of the islands and reefs in the ocean are made of coral,—the skeletons of Polyps. These islands and reefs are most abundant and most extensive in the Pacific and Indian Oceans, but the islands which skirt the coast of Florida—the Keys—are also of coral formation. Some reefs are small and have made only a little progress upward towards the surface of the water; others are miles in length and breadth, and come so near the surface of the water that it is dangerous for vessels to sail over them; and others still rise above the surface of the water forming islands which, in some cases, are covered with coral sand, and in others with a more or less luxuriant growth of tropical vegetation. Reefs stretch north and south near New Caledonia for the distance of four hundred miles, and along the northeastern coast of Australia for more than a thousand miles. When a reef or bank of coral is near the shore, it is called a *fringing reef;* when at a distance from the shore, a *barrier reef;* and when it surrounds a body of water, as is often the case in the Pacific, an *atoll* or *coral island.* The corals which form the principal part of the reefs and islands are Madrepores, Porites, Meandrinas, and Astræans; the frailer corals, such as the Sea Fans and other Gorgonias, adorn the reef as it nears the surface of the water, but do not contribute much to its growth.

From what has already been said, it is hoped that

Fig. 505. — Meandrina.

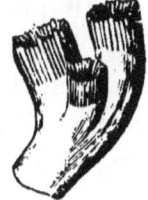

Fig. 506. — Cladocora.

Fig. 507.—Star Coral, or Astræan.

Fig. 508. — Merulina.

Fig. 509. — Oculina.

Fig. 510. — Fungus Coral.

it will be understood that the reefs and islands are not something that the Coral Polyps build, as a mason builds a house, or as a bee or wasp builds her nest or comb, but that the reefs and islands are made up of the hard parts or skeletons of Polyps that lived and died where the reef or island now stands.

Only about an inch of a growing coral mass or reef is alive, all the rest within is dead; death goes on below as fast as growth goes on above. When the reef at last grows up to the surface of the water, the Polyps die; for they cannot live out of water. The winds and waves do the rest; they break fragments from the sides of the reef and pile them nearer the center; they bring seaweeds and other floating materials, and cast them over the whole; plants at length spring up, and in the course of years the island — except its broad beaches of coral sand — is clothed with verdure, and man, perhaps, comes there and makes his home. These little Polyps, then, are increasing the amount of dry land on the surface of the globe; and in this and in other ways God makes their lives serve great and important ends.

But a history of the Polyps would be unfinished if we should not mention their connection with some of the rocks of the globe, — the limestones. It is a very interesting fact that reef corals and limestone, or marble, have essentially the same chemical composition; and it is well known that some of the coral reefs of the Pacific, which have been lifted out of water by volcanic forces, are nearly or quite as solid as ordinary marble. From these facts, and many others, geologists believe that a large part of the limestones of the globe are made out of the coral reefs that grew in the old oceans, which long before the creation of Man covered the countries where marble is now found. If this is true, many of the rocks which underlie vast countries, the marble temples and palaces of the East, the marble monuments and public buildings of our own country, the mortar upon the walls and ceilings of our houses, and the marble tables and mantelpieces so highly prized, have all come from the skeletons of these little flowerlike animals of the sea. Their skeletons have furnished even the blocks of marble which the sculptor chisels, and are thus inseparably linked with the highest department of culture and art.

SPONGES.

Naturalists formerly believed that Sponges belonged with the Protozoans. But it is now known that they are more nearly related to the Hydroids. The simplest sponges are conical or cylindrical in form, hollow,

Fig. 511.—Sponge.

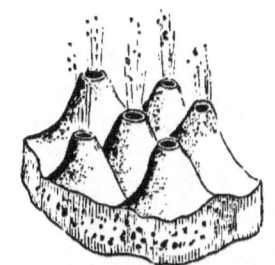

Fig. 512.—Sponge. Alive.

with a large opening at the top, while they are attached by the other end to the bottom. The wall of the Sponge is thick and is supported by a fibrous skeleton which forms the sponge of commerce. The wall is pierced by numerous holes which lead into the central cavity, sometimes directly, sometimes by means of winding passages. Water is drawn into these openings by means of cilia, and is propelled out through the central cavity to the external opening. The water brings with it minute animals and plants, and bits of dead creatures, which the cells of the Sponge seize and use as food. The Sponge branches so as to form a colony as is seen in the Figures 511 and 512. In the Bath Sponge, these branches are wholly united to each other, so that it seems a shapeless mass. Study, however, will show on the upper surface the large exhalent openings, each corresponding to a member of the Sponge colony.

The Sponges mostly live in the sea; a few inhabit the fresh water. They live in shallow water, attached to rocks, etc. Their forms are exceedingly various and often extremely beautiful. Some cover the rocks like a carpet of mosses; others grow in massive clusters; others branch like trees and shrubs; and others still take the form of the most elegant cups, goblets, and vases. They are plentiful in tropical waters about coral reefs. The Sponges of commerce come from the Red Sea, Mediterranean Sea, and West India Islands.

PROTOZOANS.

THERE is a vast number of beings so simple in their structure that naturalists were in doubt, in many cases, whether to call them plants or animals. These are now called Protozoans, a word which means *first* or *simplest animals*. A few of the forms are shown in Figures 513-521, — all much enlarged, except Figure

521. In most cases they have neither mouth nor stomach, and they are exceedingly minute and mostly microscopic. They are doubtless more numerous than all the other animals of the globe, for they live in immense numbers in every ditch and pool, every stream, pond, and lake, and almost every part of the sea. There is scarcely a drop of stagnant water that is not inhabited by some of them. They were exceedingly abundant in the past ages of the world; for their skeletons, or hard parts, fill the rocks in many places, and rocky strata hundreds of feet in thickness are wholly made up of their remains.

One group of the Protozoans is called Infusoria, from having first been found in vegetable infusions, that is, in liquids in which plants have been soaked. These are very abundant in fresh water ponds, etc. Of these, Vorticella, Figure 513, is a well-known kind.

There is another group called Rhizopods, — a word meaning *root-feet*, — because they throw out fibers or root-like appendages, as in Figures 519, 520. Many of these have a shell, and are often called Foraminifers from the pores or foramens in the shell, through which the appendages just mentioned are thrust out. The vast chalk-beds of Europe are almost wholly made of the shells of Rhizopods, which are so minute that a million are contained in a cubic inch of the chalk, so that, small as these creatures are, they have played a part in the building of the world. They live mostly in the ocean near the surface and their shells, as they die, are constantly falling to the bottom. The floor of the ocean, away from the immediate neighborhood of the land is covered with a soft mud or *ooze* mainly composed of these shells. The Nummulite, Figure

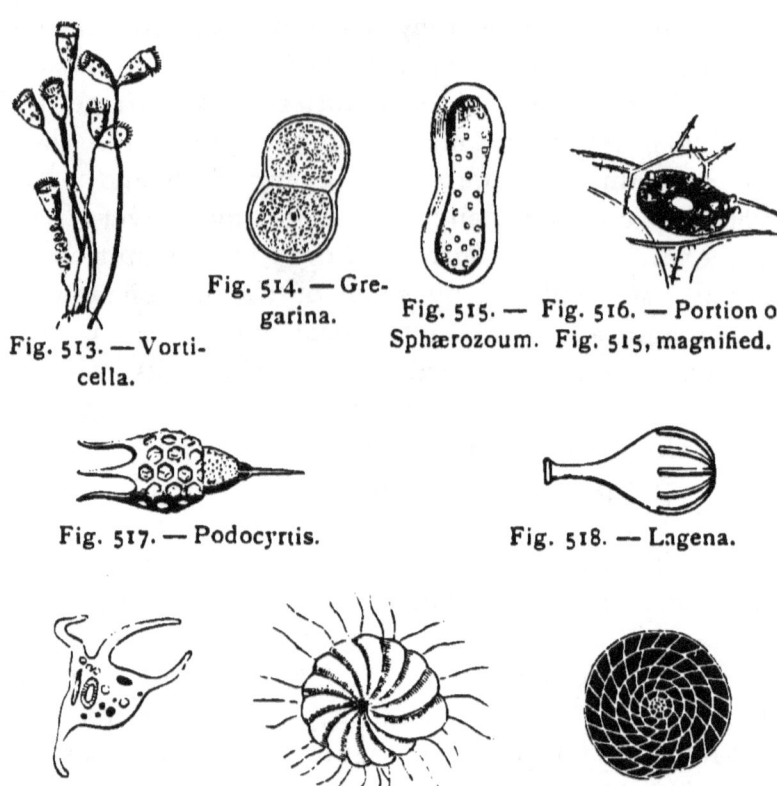

Fig. 513. — Vorticella.
Fig. 514. — Gregarina.
Fig. 515. — Fig. 516. — Portion of Sphærozoum. Fig. 515, magnified.
Fig. 517. — Podocyrtis.
Fig. 518. — Lagena.
Fig. 519 — Amœba.
Fig. 521. — Nummulite.
Fig. 520. — Polystomella.

521, is one of the Rhizopods or Foraminifers, which has a shell half an inch or more in diameter in some cases, and divided into chambers which resemble those of a Nautilus or Ammonite. Extensive beds of limestone are made of Nummulites, that of which the Pyramids of Egypt are built is filled with shells of this sort. The Amœba, Figure 519, is a Rhizopod which has no shell. It is a simple, almost fluid mass, seen only by the aid of a microscope, and it changes its form almost every moment. It has neither mouth nor stomach, yet on coming to a particle of food it readily closes around it and digests it, any part of the body being formed into mouth, stomach, or tentacles, as the occasion requires.

CONCLUSION.

In these few pages we have endeavored to make you acquainted with some of the principal forms in which animals have been created, and thus give you some idea of the Animal Kingdom. Although only a few kinds out of the many thousands now living have been mentioned, you have learned that all the animals upon our globe may be divided into eight great groups, — the Vertebrates or Backboned Animals, the Arthropods or Jointed Animals, the Mollusks or Soft-bodied Animals, the Vermes or Worms, the Echinoderms or Starfishes, the Cœlenterates or Lasso-throwers, the Sponges, and the Protozoans. The Tunicates are a small group related to the Vertebrates. It may be added that geologists tell us that all the animals of past ages, which are now known only by their remains, but which were so numerous that in many places they fill the rocks to the depth of miles, also belong to either one or the other of these groups. Naturalists call these groups Branches.

You have learned that the Vertebrates are divided into Mammals, Birds, Reptiles, Batrachians, and Fishes; that the Arthropods are divided into Insects, Arachnids, and Crustaceans; that the Mollusks are divided into Cephalopods, Gastropods, and Lamellibranchia; that the Echinoderms are divided into Sea Cucumbers, Sea

Urchins, Star fishes, Serpent Stars, and Crinoids; and that the Cœlenterates are divided into Jellyfishes and Polyps. Naturalists call these groups Classes.

You have learned that the Mammals are divided into Man, Monkeys, Carnivora or Beasts of Prey, Ungulates or Hoofed Animals, Cetaceans or Whales, Bats, Insect-eaters, Rodents or Gnawers, Edentates or Marsupials; that the Birds are divided into Birds of Prey, Climbers, Perchers, Scratchers, Runners, and Swimmers; and that the Reptiles, Batrachians, and Fishes, the Insects, Crustaceans, Cephalopods, Gastropods, and the other classes are also similarly divided into groups. Naturalists call these groups Orders.

The Orders, again, are divided into Families, — for example, the Order of Birds of Prey is divided into the Family of Vultures, the Family of Falcons and Eagles, and the Family of Owls. Families are divided into Genera, — for example, the Family of Falcons is divided into true Falcons, Hawks, Eagles, etc. Genera are divided into Species, — for example, the Genus of true Falcons is divided into the Peregrine Falcons or Duck Hawks, Pigeon Falcons or Pigeon Hawks, Sparrow Falcons or Sparrow Hawks, etc.

You have gained some idea of the way in which animals are distributed over the surface of the globe. Each zone of the earth's surface, each zone of height, each hemisphere, each grand division of the earth, has its own kinds of animals; even each of the different parts of every country has animals peculiar to itself. And it is so in the waters; each ocean and sea, each gulf and bay, and each zone of depth, has its own animal forms, such as are found nowhere else.

But the words of a book cannot fitly describe the

living beings of our globe. We need to open our eyes and study them in the world about us. We may find them everywhere, — in forest and field, on the mountain and in the sea, in every stream, pond, and lake, in every pool and ditch and bog, and in every glass of water from the spring. Every summer's day brings scores of beautiful winged forms, and on every summer's night others not less beautiful flit about our lamps, or look in at our windows, tempting us to study and admire them. And how full of interest is every living creature, whether it is the Deer bounding through the forest or over the plain, the Eagle soaring above our heads until lost amid the clouds, the Butterfly flitting from flower to flower, the Mussel plowing its way into the river's sand, or the little Polyp beneath the ocean wave. They are interesting not merely on account of their varied and beautiful forms and colors, wonderful structure, often marvelous instincts and habits, and great variety of uses, but because they are the works of God, — His thoughts expressed in visible forms. If we study these wonderful objects in the right spirit, we shall learn more of Him who made them, and who careth for them, suffering not even a sparrow to fall without His notice.

INDEX.

Acalephian Coral, 252.
Acalephs, 242.
Acicula, 222.
Actinaria, 258.
Actinia, 258, 259.
Albatross, 110.
Alcyonacea, 258.
Alcyonaria, 255.
Alligator, 117.
Alpaca, 40.
American Buffalo, 49.
Ammonite, 208, 209.
Amœba, 268.
Amphioxus, 138.
Anaconda, 119.
Angler, 128, 129.
Ant, 148.
Antedon, 242.
Antelopes, 44.
Ant-lions, 192.
Apes, 23.
Aphides, 184.
Apteryx, 103.
Arachnids, 193.
Argonauts, 206, 207.
Armadillos, 69.
Arthropods, 139.
Asilus Fly, 169, 172.
Ass, 51.
Astarte, 228.
Asterias, 156.
Astræans, 260, 263.
Astroides, 261.
Astrophyton, 240.
Atlanta, 222.
Auks, 114.
Aurelia, 247.
Avicula, 225, 227.
Axolotl, 123.
Aye-aye, 26.

Baboons, 25.
Badger, 34, 36.
Bald Eagle, 77, 78.
Barnacles, 201.
Basket Fish, 240.
Bass, 125, 126.
Bats, 57.
Batrachians, 120.
Bears, 35, 36.
Beautiful Deïopeia, 162.
Beaver, 65.
Bee Fly, 169, 172.
Bees, 143.
Beetles, 172.
Big-horn Sheep, 47, 48.
Birds, 72.
Birds of Prey, 75.
Bison, 49.
Bittern, 105.
Blackbirds, 98.
Black Rat, 66.
Black Snake, 119, 120.
Black Snowbird, 96.
Black Woodcock, 83.
Blindfish, 130, 131.
Bluebird, 89.
Bluefish, 127, 128.
Blue Jay, 99.
Boa, 119.
Bobolink, 97.
Bolina, 245.
Borers, 180.
Botflies, 169, 172.
Box Turtle, 116.
Brachiopods, 231.
Brain Coral, 260.
Bream, 125, 126.
Bryozoa, 232.
Buccinum, 214.
Buffalo, 49.

Bugs, 181.
Bulla, 222.
Bullimus, 221.
Bunting, 96.
Buprestidans, 178.
Buprestis, 178.
Burbot, 132.
Butcher Bird, 91.
Butterflies, 152.

Caddice Fly, 193.
California Vulture, 75, 76.
Camel Bird, 103.
Camels, 39.
Campanularian, 249, 251.
Cankerworm, 166.
Canvasback Duck, 109.
Capricorn Beetles, 180.
Capybara, 65.
Carabidæ, 174.
Cardicum, 225.
Caribou, 42.
Carnivora, 27.
Carp, 128.
Carrion Beetles, 175.
Cashmere Goat, 48.
Cassowaries, 103.
Catbird, 93.
Caterpillar Hunter, 174.
Cats, 27.
Cecropia, 163.
Centipedes, 197.
Cephalopods, 205.
Cerithiums, 216, 217.
Cetaceans, 53.
Chamois, 46, 47.
Cheiroptera, 57.
Chewink, 97.
Chickadees, 94.
Chimpanzee, 23, 25.
Chipmunk, 63.
Chiton, 219.
Chrysalis, 142.
Chrysalis Shell, 221.
Chrysomela, 181.
Chuck-will's Widow, 85.
Cicadas, 181.

Civets, 31.
Clams, 227, 229.
Climbers, 80.
Clio, 223.
Clothes' Moth, 168.
Cockroaches, 185, 186.
Cocoon, 142.
Cod, 130, 132.
Cœlenterates, 242.
Cœnopsammia, 261.
Coleoptera, 172.
Conches, 211.
Condor, 76.
Cones, 214, 215.
Congo Snake, 123.
Conner, 128.
Coral, 254, 256.
Corydalus, 190.
Coryne, 248, 249.
Cowbird, 98.
Cowries, 215, 216.
Crabs, 197.
Cranes, 103.
Crawfish, 201.
Creepers, 93.
Crepidula, 219.
Crickets, 187.
Crinoids, 241.
Crocodile, 117.
Crossbills, 95.
Crow, 99.
Crustaceans, 197.
Ctenophora, 244.
Cuckoos, 81.
Cucumber Beetle, 181.
Cud-chewers, 39.
Curculios, 179.
Curlews, 103, 107.
Cursores, 103.
Cuttlefish, 209.
Cyclostoma, 222.
Cytherea, 228.

Dace, 128.
Darning Needle, 190.
Darter, 125.
Deer, 40.

Diptera, 168.
Discophora, 246.
Diver, 112.
Dog-headed Monkeys, 26.
Dogs, 29.
Dolphin, 56, 128, 129.
Doris, 222.
Doves, 100.
Dragon Fly, 190.
Duckbill, 71.
Duck Hawk, 78.
Ducks, 109.

Eagle, 78.
Ear-shell, 219.
Earthworms, 230.
Earwig, 185.
Echinarachnius, 236.
Echinoderms, 233.
Echinoidea, 234.
Edentates, 69.
Eel, 132, 133.
Eelpouts, 128, 129.
Eider Duck, 109.
Elaters, 178.
Elephants, 52.
Elk, 43.
Elysia, 222.
Eolis, 222.
Ephemera, 189.
Ephyra, 247.

Falcons, 76, 78.
Feather Star, 242.
Field Mouse, 67.
Finches, 95.
Firefly, 178.
Fisher, 31.
Fishes, 124.
Fishhawks, 78.
Flesh-eaters, 27.
Flounders, 132, 133.
Flower Beetles, 177.
Fly, 168.
Flycatchers, 87.
Flying Fishes, 130, 131.
Flying Squirrel, 62.

Foraminifers, 268.
Foxes, 30.
Frogs, 121.
Frog Shell, 213.
Fungus Corals, 260, 263.
Fusus, 214, 215.

Galeopithecus, 58.
Galley Worm, 197.
Gallflies, 150.
Gallinules, 103, 107.
Garfish, 130, 131.
Garpike, 136.
Gastrochæna, 229.
Gastropods, 210.
Gazelle, 46.
Geese, 108.
Geometers, 166.
Gnawers, 60.
Goats, 48.
Godwits, 103, 107.
Goldfinch, 95.
Goldsmith Beetle, 177.
Goosefish, 128, 129.
Gophers, 63, 114.
Gordius, 233.
Gorgonias, 257.
Gorilla, 25.
Grallatores, 103.
Grasshoppers, 188.
Grebes, 112.
Green Turtle, 115.
Greenhead, 109.
Greenland Whale, 53
Gregarina, 268.
Grizzly Bear, 36.
Grosbeaks, 96.
Ground Beetles, 174, 177.
Ground Robin, 96.
Grouse, 101.
Gulls, 111.

Hagfish, 136, 138.
Hair Worm, 233.
Halibut, 133.
Hammerhead Shark, 137.
Hares, 68.

Harp Shell, 214, 215.
Harvest Flies, 181.
Harvest Mouse, 66.
Hawk Moths, 158.
Hawks, 76.
Hedgehogs, 60.
Helicina, 222.
Helix, 220, 221.
Hemiptera, 181.
Hens, 100.
Hermit Crab, 198.
Hermit Thrush, 88.
Herons, 103.
Herring, 130, 131.
Hessian Fly, 169.
Heteropods, 222.
Hipparchians, 157.
Hippopotamus, 39.
Hoary Bat, 58.
Hogs, 38.
Holothurians, 233.
Horn Bug, 176.
Horned Pout, 130, 131.
Horned Toad, 118.
Hornet, 147.
Horse, 50.
Horsefly, 169, 171.
Horseshoe Crab, 202.
Humble Bee, 146.
Humming Birds, 84.
Hydroids, 248.
Hyena, 29.
Hylea, 223.
Hymenoptera, 143.

Ibises, 105.
Ichneumon, 149.
Idyia, 245.
Imago, 141, 142.
Infusoria, 267.
Insect-eaters, or Insectivora, 58.
Insects, 139.
Insessores, 84.
Io, 217.

Jaguar, 29.

Jay, 99.
Jellyfishes, 242.
Jumping Mouse, 67.

Kahau, 25.
Kangaroos, 70.
Katydid, 188.
Kingbird, 87.
Kingfishers, 86.

Lacewings, 192.
Ladybird, 181.
Lagena, 268.
Lamellibranchia, 223,
Lammergeyer, 76.
Lampreys, 136, 138.
Lancelet, 138.
Land Snails, 220, 221.
Larks, 98.
Larva, 141, 142.
Lasso-throwers, 242.
Leaf Rollers, 167.
Leda, 225.
Lemur, 26.
Leopard, 27.
Leopard Frog, 121.
Lepidoptera, 152.
Limacina, 223.
Limnæa, 221.
Limnæidæ, 220, 221.
Limpets, 219, 220.
Lingula, 232.
Lion, 27.
Lizards, 117.
Llamas, 39, 40.
Lobsters, 199.
Locusts, 187.
Long-horned Beetles, 180.
Loon, 112.
Lumpfish, 132, 133.
Luna Moth, 163, 165.
Lycosa, 194.
Lynx, 28, 29.

Mackerel, 126, 127.
Mactra, 228.
Madreporaria, 258.

INDEX. 277

Madrepores, 259, 261.
Magpie, 99.
Maki, 26.
Mallard, 109.
Mammals, 22.
Man, 22.
Mandrills, 26.
Mantis, 186.
Marginella, 215, 216.
Marmoset, 26.
Marsupials, 69.
Martens, 31.
Martin, 91.
Maryland Yellowthroat, 90.
Mastodon, 53.
May Flies, 189.
Meadow Lark, 98.
Meandrinas, 260, 263.
Medusæ, 246.
Melanias, 217, 218.
Mellita, 237.
Merulina, 263.
Mice, 66.
Mink, 32.
Misippus Butterfly, 157.
Miter Shell, 215, 216.
Mocking Birds, 92.
Moles, 59.
Mollusks, 203.
Monkeys, 23.
Monotremes, 71.
Moose, 41.
Mosquito, 168.
Mother Cary's Chickens, 111.
Moths, 158.
Mountain Sheep, 47, 48.
Mud Puppy, 123.
Mullets, 128, 129.
Murex Shells, 212, 213.
Musk Deer, 44.
Musk Ox, 48.
Muskrat, 67.
Mussels, 227.
Myriapods, 197.
Mytilus, 225.
Myxine, 136, 138.

Natatores, 108.
Naticas, 215, 216.
Nautili, 208, 209.
Necturus, 123.
Nerita, 219.
Neritina, 219.
Net-winged Insects, 191.
Neuroptera, 191.
Nighthawk, 85.
Nightingale, 89, 90.
Nummulite, 267.
Nuthatches, 93.
Nymphalis Butterflies, 157.

Octopus, 206, 207.
Oculina, 260, 263.
Odd-toed Ungulates, 50.
Oliva Shell, 214, 215.
Ophiurans, 240.
Opossums, 70.
Orang-outang, 24, 25.
Organ-pipe Coral, 256, 258.
Oriole, 99.
Orthoptera, 185.
Osprey, 78.
Ostriches, 103.
Otters, 33, 34.
Ouzel, 89.
Owls, 79.
Oxen, 50.
Oyster, 225.

Paludina, 217.
Pandora, 229.
Panther, 27.
Papilio Butterflies, 155.
Parasitic Worms, 233.
Parrots, 81.
Partridge, 101, 103.
Patella, 219.
Peach-tree Borer, 161.
Pearl Oyster, 227.
Peccaries, 38.
Pectens, 225, 227.
Penguins, 113.
Perch, 125, 126.

Perchers, 84
Peregrine Falcon, 78.
Periwinkles, 217, 218.
Petrels, 111.
Pewees, 87.
Phanæus, 177.
Philodice, 156.
Phœbe Bird, 87.
Pholades, 228, 229.
Physa, 221.
Pickerel, 128, 131.
Pigeons, 100.
Pike, 128.
Pilot Fish, 127.
Pine Marten, 31.
Pipefishes, 134, 135.
Planorbis, 221.
Plant Lice, 184.
Planula, 246.
Platypus, 71.
Plectognathi, 134.
Pleurobrachia, 244.
Plovers, 103, 106.
Pocket Gopher, 64.
Podocyrtis, 268.
Polyps, 252.
Polyphemus Moth, 163, 165.
Polystomella, 268.
Polyzoa, 232.
Pond Snails, 220, 221.
Porcupines, 67.
Porgee, 126.
Porites, 259, 261.
Porpoise, 56.
Portugese Man-of-war, 250, 251.
Poulp, 206, 207.
Prairie Chicken, 101.
Prairie Dog, 63.
Primnoa, 256, 257.
Prionus, 181.
Proboscidea, 52.
Promethea, 163, 164.
Pronghorn Antelope, 45.
Protozoans, 266.
Pseudoneuroptera, 189.
Ptarmigans, 101.

Pteropods, 222.
Puffers, 134, 135.
Puffin, 114.
Puma, 28.
Pupa, 141, 142.
Pupa Shell, 221.
Pyramid Shells, 215, 216.
Pyrula, 213.
Python, 119.

Quadrumana, 23.
Quails, 102.

Rabbits, 68.
Raccoon, 36.
Rails, 103, 107.
Raptores, 75.
Rasores, 100.
Rattlesnake, 120.
Rats, 66.
Raven, 99.
Rays, 136, 138.
Razor Shell, 227, 228.
Red Bat, 58.
Red Coral, 257.
Red Squirrel, 61.
Reedbirds, 97.
Reindeer, 42.
Remora, 132, 133.
Renilla, 256, 257.
Reptiles, 114.
Rhea, 103.
Rhinoceros, 51.
Rhizopods, 267.
Ricinula, 214.
Right Whale, 53.
River Mussels, 226, 227.
River Snails, 217, 218.
Robins, 88.
Rocky Mountain Goat, 45, 46.
Rodents, 60.
Rotula, 237.
Rove Beetle, 176.
Ruby-crowned Wren, 88, 89.
Ruffed Grouse, 101.
Ruminants, 39.
Runners, 103.

INDEX.

Sables. 31.
Salamanders, 122.
Salmon, 130, 131.
Salt-marsh Moth, 162.
Sand Fleas, 201.
Sapsuckers, 83.
Satyrus Butterfly, 157.
Sawfish, 137.
Sawflies, 151.
Scallops, 225, 227.
Scansores, 80.
Scarabæidæ, 176.
Scorpion Bug, 184.
Scorpions, 196.
Scorpion Shell, 212.
Scratchers, 100.
Sculpins, 126, 127.
Scupaug, 127.
Scutularians, 251.
Scyphostoma, 246.
Sea Anemones, 252, 258.
Sea Cucumbers, 233.
Sea Fan, 256, 257.
Sea Horse, 134, 135.
Seal, 37.
Sea-pen, 257.
Sea Ravens, 125, 126.
Sea Robins, 125, 126.
Sea Slugs, 222.
Sea Stars, 238.
Sea Urchins, 234.
Seaworm, 230.
Selachians, 136.
Serpents, 118.
Serpent Stars, 240.
Serpula, 230.
Sesias, 161.
Sharks, 136, 137, 138.
Sheep, 47.
Shiners, 128, 131.
Shipworms, 228, 229.
Shore Lark, 94.
Shrews, 58.
Shrike, 91.
Shrimp, 198, 199.
Sigaretus, 216.
Silkworm Moths, 161.

Siphonophora, 251.
Siren, 123.
Skates, 136, 138.
Skippers, 158.
Skunk, 34.
Skylark, 94.
Slugs, 221, 222.
Snails, 210.
Snakes, 118.
Snap Beetles, 178.
Snapping Turtle, 114.
Snipes, 103, 106.
Snowbird, 96.
Soft-finned Fishes, 128.
Solen, 227, 228.
Sparrows, 96.
Sparrow Hawk, 78.
Sperm Whale, 54.
Sphærium, 228.
Sphærozoum, 268.
Sphargis, 116.
Sphingidæ, 158, 161.
Sphinx, 160.
Spider Monkey, 25, 26.
Spiders, 193.
Spine-finned Fishes, 126.
Spiny Ant-eater, 72.
Spirula, 208, 209.
Sponges, 265.
Spotted Pelidnota, 177.
Spring Beetles, 178.
Squash Bug, 185.
Squid, 206, 207.
Squirrels, 61.
Stake-driver, 105.
Starfishes, 238.
Stargazers, 125, 126.
Stickleback, 125, 126.
Stilts, 103, 107.
Stone Fly, 190.
Stormy Petrel, 111.
Straight-winged Insects, 185.
Striped Gopher, 63.
Strobila, 246.
Strombs, 211, 212.
Sturgeon, 134.
Suckers, 128, 136.

INDEX.

Sunfishes, 134, 135.
Surgeon, 128, 129.
Swallows, 90.
Swans, 108.
Swimmers, 108.
Sword Bearers, 188.
Swordfish, 126, 127.

Tapeworm, 233.
Tapirs, 52.
Tellina, 228.
Tent Caterpillar, 165.
Terebratula, 232.
Terns, 111.
Threadworms, 233.
Thrushes, 88.
Thyasira, 228.
Tiaropsis, 249, 251.
Tiger, 28.
Tiger Beetles, 173.
Tinean, 168.
Titmouse, 94.
Toadfish, 128.
Toads, 121.
Toothless Mammals, 69.
Toothshell, 219.
Tornatella, 222.
Torpedo, 136, 138.
Tortoises, 114.
Tower Shell, 217, 218.
Towhee Bunting, 96.
Tree Beetles, 177.
Tree Frog, 122.
Tree Hoppers, 183.
Trichina, 233.
Tridacna, 225.
Trilobite, 201.
Triton, 122.
Tritonia, 222.
Tritonium, 213.
Trivia, 216.
Trochus, 219.
Trout, 130, 131.
Trunkfish, 134, 135.
Tubularia, 249, 251.
Tuft-gilled Fishes, 133.
Tunicates, 139.

Turkeys, 100.
Turnstones, 103.
Turritella, 217, 218.
Turtles, 114.
Two-winged Insects, 168.
Ungulates, 37.
Unios, 226, 227.

Valvata, 217.
Vermes, 229.
Vermetus, 217, 218.
Verrucella, 256.
Vertebrates, 19.
Violet Snails, 218, 219.
Vireos, 92.
Virginia Deer, 43, 44.
Volutes, 215, 216.
Vorticella, 267, 268.
Vultures, 76.

Waders, 103.
Walking Leaf, 186.
Walking Stick, 186.
Walrus, 37.
Wapiti, 43.
Warblers, 90.
Wasps, 146.
Water Beetles, 175.
Watering-pot Shell, 229.
Weakfish, 126, 127.
Weasels, 31, 32.
Weevils, 179.
Wentletraps, 217, 218.
Whales, 53.
Wharf Rat, 66.
Wheat Fly, 170.
Whelks, 213.
Whippoorwill, 85.
Whirligig Beetle, 175.
White Boar, 38.
White Butterflies, 156.
White-footed Mouse, 66, 67.
White Whale, 57.
Wild Cat, 29.
Wing-shells, 211.
Wolverine, 33.
Wolves, 29, 30.

Wombat, 71.
Woodcocks, 103.
Wood Duck, 109.
Woodpeckers, 82.
Worms, 229.
Worm-shell, 217, 218.
Wrens, 93.

Yellowbird, 95.
Yellow Butterflies, 156.
Yellowlegs, 103, 106.
Yellowthroat, 90.

Zebra, 51.
Zebu, 50.

Zoology and Natural History.

Cooper's Animal Life.
By Sarah Cooper. . . . . $1.25.

Animal life in the sea and on the land. A zoology for young people. Especial attention has been given to the structure of animals, and to the wonderful adaptation of this structure to their habits of life.

Holder's Elementary Zoology.
By C. F. Holder. . . . . $1.20.

A text-book designed to present in concise language the life-histories of the groups that constitute the animal kingdom, giving special prominence to distinctive characteristics and habits.

Hooker's Child's Book of Nature.
Part II. Animals. By Worthington Hooker, M. D. $0.44.

While this work is well suited as a class-book for schools, its fresh and simple style can not fail to render it a great favorite for family reading.

Hooker's Natural History.
By Worthington Hooker, M. D. . . . . $0.90.

For the use of schools and families. Illustrated by three hundred engravings. The book includes only that which every well-informed person ought to know, and excludes all which is of interest only to those who intend to be thorough zoologists.

Morse's First Book in Zoology.
By E. S. Morse, Ph. D. . . . . $0.87.

Prepared for the use of pupils who wish to gain a general knowledge concerning the common animals of the country. The examples presented for study are such as are common and familiar to every school-boy.

Nicholson's Text-Book of Zoology.
By H. A. Nicholson, M. D. . . . . $1.38.

Revised edition. A work strictly elementary, designed for junior students. Illustrated with numerous engravings. It contains an Appendix, Glossary, and Index.

Steele's New Popular Zoology.
By J. Dorman Steele, Ph. D. . . . . $1.20.

This book proceeds, by natural development, from the lowest form of organism to man. A cut is given of every animal named, since a good picture of an object is worth more than pages of description.

Tenney's Elements of Zoology.
By Sanborn Tenney, A. M. . . . . $1.60.

Illustrated by seven hundred and fifty wood engravings. It gives an outline of the animal kingdom, and presents the elementary facts and principles of zoology.

Tenney's Natural History of Animals.
By Sanborn Tenney and Abby A. Tenney. . . . $1.20.

A brief account of the animal kingdom, for the use of parents and teachers. Illustrated by five hundred wood engravings, chiefly of North American animals.

Copies mailed, post-paid, on receipt of price. Complete price-list sent on application.

AMERICAN BOOK COMPANY,
NEW YORK ∴ CINCINNATI ∴ CHICAGO.

PUBLICATIONS OF THE AMERICAN BOOK COMPANY.

PHYSIOLOGY.

ECLECTIC PHYSIOLOGY AND HYGIENE. By ELI F. BROWN, M.D. 12mo, cloth, 189 pages . . . 60 cents

A low-priced text-book, adapted to the requirements of the new school laws relating to instruction in this branch.

The effects of narcotics and stimulants on the body and mind are duly set forth. Much attention given to proper sanitary condition in the home. Language plain and didactic in style.

FOSTER & TRACY'S PHYSIOLOGY AND HYGIENE.

(Science Primer Series.) By M. FOSTER, M.D., and R. S. TRACY, M.D. 18mo, cloth, 170 pages . . . . . 35 cents

Foster's valuable work on Physiology is supplemented by the Chapters on Hygiene by Tracy. This Primer is an attempt to explain in the most simple manner possible some of the most important and general facts of physiology. The whole field of physiology is not taken up, but the *fundamental truths* are carefully set forth.

HUNT'S PRINCIPLES OF HYGIENE. By EZRA M. HUNT, M.D. 12mo, cloth, illustrated, 400 pages . . 90 cents

This is an authoritative work on an original plan which makes the knowledge of hygiene and the practice of its principles the first aim, using the study of anatomy and physiology as a means to this end and not the end itself.

The effects of alcoholic stimulants and narcotics are treated in proper connection. No doubtful views are included.

JARVIS'S PHYSIOLOGY AND LAWS OF HEALTH.

By EDWARD JARVIS, M.D. 12mo, cloth, 427 pages . $1.20

A work that approaches the subject with a proper view of the true object in teaching physiology in schools, viz., that scholars may know how to take care of their own health. It makes the *science* a secondary consideration, and only so far as is necessary for the comprehension of the laws of health.

HOW WE LIVE; or, The Human Body and How to Take Care of It. By JAMES JOHONNOT, EUGENE BOUTON, Ph.D., and HENRY D. DIDAMA, M.D. . . . . 40 cents

An Elementary Course in Anatomy, Physiology, and Hygiene. A text-book thoroughly adapted to elementary instruction in the public schools; giving special attention to the laws of hygiene (including the effects of alcohol and narcotics upon the human system), as ascertained from a careful study of anatomy and physiology; containing also a full Glossary of Terms, complete Index, etc.

KELLOGG'S FIRST BOOK IN PHYSIOLOGY AND HYGIENE. By J. H. KELLOGG, M.D., 170 pages, 40 cents

The design of this book is to present in as simple a manner as possible such hygienic and physiological facts as are necessary to acquaint children with the laws of healthful living. Technical terms have been avoided, and no matter has been introduced which has not some practical reference to the preservation of health.

[*48]

PHYSIOLOGY—Continued.

SMITH'S PRIMER OF PHYSIOLOGY AND HYGIENE.
By William Thayer Smith, M.D. Cloth, illustrated, 124 pages . . . . 30 cents

Designed for children. It treats the subject in a novel and interesting manner, and shows the effects of stimulants and narcotics on each part or function of the body as studied.

SMITH'S ELEMENTARY PHYSIOLOGY AND HYGIENE.
By William Thayer Smith, M.D. Full cloth, illustrated, about 200 pages . . . . . 50 cents

This book has been prepared with great care to meet the increasing demand for a manual of suitable style and compass for ungraded and grammar schools, giving special attention to the care of the body and the preservation of health, and inculcating practical temperance by showing the injurious effects of alcoholic stimulants and narcotics.

STEELE'S HYGIENIC PHYSIOLOGY.
By J. Dorman Steele, Ph.D. 12mo, cloth, 276 pages . . . $1.00

With especial reference to alcoholic drinks and narcotics. Throughout the book there are given, in text and foot-note, experiments that can be performed by teacher or pupil, and which will induce some easy dissections to be made in every class. Great attention is given to the subject of ventilation, which is now attracting so much attention throughout the country.

STEELE'S ABRIDGED HYGIENIC PHYSIOLOGY.
By J. Dorman Steele, Ph.D. 12mo, cloth, 192 pages, 50 cents

This work embraces many of the same excellent features as the above. Adapted for younger pupils.

THE ESSENTIALS OF ANATOMY, PHYSIOLOGY, AND HYGIENE.
By Roger S. Tracy, M.D. 12mo, 299 pages $1.00

A clear and intelligible account of the structures, activities and care of the human system. Great prominence is given to anatomical and physiological facts, which are necessary preliminaries to instruction in hygiene.

HEALTH LESSONS.
By Jerome Walker, M.D. 12mo, 194 pages . . . . 48 cents

The object of this attractive little work is to teach health subjects to children in an interesting and impressive way, and to present to their minds hygienic facts that are easily comprehended and that quickly lead to practical results in their daily habits and observations.

Copies of these or any of the publications of the American Book Company for the use of teachers or school officers, or for examination with a view to introduction, will be sent by mail, postpaid, on receipt of the list or introduction price.

AMERICAN BOOK COMPANY,
NEW YORK ∴ CINCINNATI ∴ CHICAGO.

Physics and Natural Philosophy.

Arnott's Physics. Seventh edition. Edited by ALEXANDER BAIN, LL. D. . . . . $2.40.
An excellent and complete treatise on Natural Philosophy and Astronomy.

Cooley's New Text-Book of Physics. . . 90 cents.
An elementary course in Natural Philosophy designed for high schools and academies. It gives special prominence to the principle of energy.

Cooley's Easy Experiments. . . . . 52 cents.
A course of experiments in physical science for oral instruction in common schools. The experiments are such as intelligent boys and girls can make with little assistance.

Cooley's Elements of Natural Philosophy. 72 cents.
Revised edition. For common and high schools. It presents the most elementary facts of Natural Philosophy in an entertaining and instructive manner.

Everett's Outlines of Natural Philosophy. 84 cents.
Designed for schools and for general readers. It is at once easy enough for a class reading-book, and precise enough for a text-book.

Gillet and Rolfe's First Book in Natural Philosophy. . . . . 60 cents.
For the use of schools and academies. It contains a brief, simple, and natural statement of the facts and principles of the science.

Hooker's Child's Book of Nature. Part III. 44 cents.
A course of lessons on Air, Water, Heat, Light, etc., admirably illustrated.

Norton's Elements of Physics. . . . . . 80 cents.
A systematic epitome of the science, designed as a text-book for academies and common schools.

Norton's Natural Philosophy. . . . . $1.10.
A selection of the facts and principles of Natural Philosophy, but adapted to the requirements of the pupil.

Peck's Ganot Revised. . . . . $1.20.
An introductory course in Natural Philosophy for the use of high schools and academies, edited from Ganot's Popular Physics by William G. Peck.

Quackenbos's Natural Philosophy. . . . . $1.22.
Designed to exhibit the application of scientific principles in every-day life.

Steele's Popular Physics. . . . . $1.00.
Designed both to interest and instruct the pupil. Written in simple language, and containing many practical illustrations.

Trowbridge's New Physics. . . . . $1.20.
A manual of experimental study for high schools and preparatory schools.

Wells's Natural Philosophy. . . . . $1.15.
This embodies the latest and best results of scientific discovery and research.

Copies mailed, post-paid, on receipt of price. Complete price-list sent on application.

AMERICAN BOOK COMPANY.

NEW YORK ·:· CINCINNATI ·:· CHICAGO

Chemistry.

Armstrong and Norton's Laboratory Manual of Chemistry.
By JAMES E. ARMSTRONG and JAMES H. NORTON. . . . 50 cents.
One hundred and sixty-four experiments, or a year's work, clearly though briefly explained, and employing simple and for the most part inexpensive apparatus.

Brewster's First Book of Chemistry.
By MARY-SHAW BREWSTER. . . . . 66 cents.
A course of experiments of the most elementary character for the guidance of children in the simplest preliminary chemical operations. The simplest apparatus is employed.

Clarke's Elements of Chemistry.
By F. W. CLARKE. . . . . $1.20.
A class-book intended to serve, not only as a complete course for pupils studying chemistry merely as part of a general education, but also as a scientific basis for subsequent higher study.

Cooley's New Elementary Chemistry for Beginners.
By LE ROY C. COOLEY. . . . . 72 cents.
This is emphatically a book of experimental chemistry. Facts and principles are derived from experiments, and are clearly stated in their order.

Cooley's New Text-Book of Chemistry.
By LE ROY C. COOLEY. . . . . 90 cents.
A text-book of chemistry for use in high schools and academies.

Eliot and Storer's Elementary Chemistry.
Abridged from Eliot and Storer's Manual, by WILLIAM RIPLEY NICHOLS, with the co-operation of the authors. . . . . . $1.08.
Adapted for use in high schools, normal schools, and colleges.

Steele's New Popular Chemistry.
By J. DORMAN STEELE, Ph. D. . . . . $1.00.
Devoted to principles and practical applications. Not a work of reference, but a pleasant study. Only the main facts and principles of the science are given.

Stoddard's Qualitative Analysis.
By JOHN T. STODDARD, Ph. D. . . . . 75 cents.
An outline of qualitative analysis for beginners. The student is expected to make the reactions and express them in written equations.

Stoddard's Lecture Notes on General Chemistry.
Part I. Non-Metals. . . . . $0.75.
Part II. Metals. . . . . 1.00.
Designed as a basis of notes to be taken on a first course of experimental lectures on general chemistry, to relieve the student from the most irksome part of his note-taking.

Youmans's Class-Book of Chemistry.
By EDWARD L. YOUMANS, M. D. Third edition. Revised and partly rewritten by WILLIAM J. YOUMANS, M. D. . . . . $1.22.
Designed as a popular introduction to the study of the science, for schools, colleges, and general reading. With a colored frontispiece and 158 illustrations.

Copies mailed, post-paid, on receipt of price. Full price-list sent on application.

AMERICAN BOOK COMPANY,
NEW YORK .:. CINCINNATI .:. CHICAGO.

Geology.

Andrews's Elementary Geology.
By E. B. ANDREWS, LL. D. . . . . $1.00.

This book is designed for students and readers of the Interior States, and therefore has its chief references to home geology. The scope is limited, to adapt it to beginners.

Dana's Geological Story Briefly Told.
By JAMES D. DANA, LL. D. . . . . $1.15.

With numerous illustrations. An introduction to geology for the general reader, and for beginners in the science. It contains a complete alphabetical index of subjects.

Dana's Manual of Geology.
By JAMES D. DANA, LL. D. . . . . $3.84.

This is a treatise on the principles of the science adapted to the wants of the American student, with special reference to American geological history. The illustrations are numerous, accurate, and well executed.

Dana's New Text-Book of Geology.
By JAMES D. DANA, LL. D. . . . . $2.00.

On the plan of the Manual, designed for schools and academies. The explanations are simple, and at the same time complete.

Le Conte's Compend of Geology.
By JOSEPH LE CONTE. . . . . $1.20.

A book designed to interest the pupil, and to convey real scientific knowledge. It cultivates the habit of observation by directing the attention of the pupil to scientific phenomena.

Nicholson's Text-Book of Geology.
By H. A. NICHOLSON. . . . . $1.05.

This presents the leading principles and facts of geological science within as brief a compass as is compatible with clearness and accuracy.

Steele's Fourteen Weeks in Geology.
By J. DORMAN STEELE, Ph. D. . . . . $1.00.

Designed to make science interesting by omitting those details which are valuable only to the scientific man, and by presenting only those points of general importance with which every well-informed person wishes to be acquainted.

Williams's Applied Geology.
By S. G. WILLIAMS. . . . . $1.20.

A treatise on the industrial relations of geological structure, and on the nature, occurrence, and uses of substances derived from geological sources. It gives a connected and systematic view of the applications of geology to the various uses of mankind.

Copies mailed, post-paid, on receipt of price. Full price-list sent on application.

AMERICAN BOOK COMPANY,
NEW YORK .:. CINCINNATI .:. CHICAGO.

www.ingramcontent.com/pod-product-compliance
Lightning Source LLC
Chambersburg PA
CBHW031341230426
43670CB00006B/402